Chemical Demonstrations

Chemical Demonstrations
A Sourcebook for Teachers

Lee R. Summerlin
The University of Alabama at Birmingham

James L. Ealy, Jr.
The Hill School

AMERICAN CHEMICAL SOCIETY
WASHINGTON, DC. 1985

Library of Congress Cataloging in Publication Data

Summerlin, Lee R.
Chemical demonstrations.

Includes index.

1. Chemistry—Experiments. I. Ealy, Jim, 1943–
II. Title.

QD43.S77 1985 542 85-11240

ISBN 0-8412-0923-5

Book cover design and illustrations: Pamela Lewis
Copy Editor: Susan Robinson
Production Editor: Hilary Kanter
Managing Editor: Janet S. Dodd
Text typesetting: The Lodders of Rockville, Rockville, Md.
Front matter and index typesetting: Hot Type Ltd., Washington, D.C.
Typeface: Melior
Printing and binding: R. R. Donnelley and Sons Company, Harrisonburg, Va.

PRINTED IN THE UNITED STATES OF AMERICA

Second Printing 1986

About the Authors

LEE R. SUMMERLIN received his B.A. in Chemistry/Biology from Samford University, his M.S. in Biology/Chemistry from Birmingham-Southern College, and his Ph.D. in Chemistry/Science Education from the University of Maryland. He has held many teaching and administrative positions as well as served as a consultant to various agencies and organizations, companies, colleges and universities, and school boards in this country and abroad. He has served as a peer review panelist for several science educational development programs and as a chemistry consultant to various National Science Foundation Institutes. He has presented many seminars and published a number of books on methods and aspects of teaching chemistry and science. He has also conducted numerous workshops on chemical demonstrations throughout the country. He served as coordinator of the Institute for Chemical Education program at the University of California, Berkeley, in the summer of 1986. Dr. Summerlin has held office or had major committee assignments in the American Chemical Society, the National Science Teachers Association, the American Association for the Advancement of Science, the Florida Association of Science Teachers, and the Association for the Education of Teachers of Science. He was the recipient of the American Chemical Society Outstanding Chemistry Teacher Award (1967), the James B. Conant Award (1969), and the Du Pont Award (1970). He received the 1985 Ingalls Award, which is given to the Outstanding Teacher at the University of Alabama at Birmingham, and was a national recipient of the 1986 Chemical Manufacturers Association Catalyst Award.

JAMES LEE EALY, JR., received his B.A. in Chemistry/Biology from Shippensburg University. Since that time he has worked at the Mercersburg Academy (Chemistry and Biology Master), The Greensboro Day School (Science Department Chairman), and The Hill School (Head, Chemistry Department). In each of these positions, he worked to develop demonstrations as well as courses that would teach students how to learn the "scientific method" as well as the basics of chemistry. He developed a laboratory manual for use at The Hill School and is a teacher in the Chemistry Workshop presented by The Taft School in Watertown, Connecticut. In addition to working at The Hill School, Mr. Ealy serves as a marine consultant; his latest project was to develop nine training manuals to be used by a consortium of fourteen oil companies in Alaska. Mr. Ealy has also served on various committees including the American Chemical Society Committee on Advanced Chemistry Exam Writing and the Test Writing Committee for the 1985 Chemistry Olympiad. He serves as a subject editor for the *Journal of Chemical Education* and has presented demonstrations around the country and papers at several ChemEd International Meetings.

A Word of Caution

Demonstrations can be fun, and they add excitement to the teaching of chemistry. However, this volume is intended for use by professional chemists and chemistry teachers. Even the simplest demonstration is potentially dangerous when performed by someone lacking the manipulative skills and knowledge of chemistry necessary to understand the reactions involved. Every precaution must be taken to ensure the safety of the demonstrator and the students. You should follow the directions given for each demonstration and not exceed the recommended amounts of chemicals.

Contents

Preface **ix**

Properties of Atoms

1. Electronegativity, Atomic Diameter, and Ionization Energy **3**

Gases

2. Gas Densities **7**
3. Properties of Gases: Pressure and Suction **8**
4. Temperature and Pressure Relationships **9**
5. Solubility of a Gas: The Ammonia Fountain **10**
6. Preparation of Oxygen Gas from Laundry Bleach **12**
7. Preparation of Chlorine Gas from Laundry Bleach **13**
8. Diffusion of Gases **14**
9. Production of a Gas: Acetylene **16**
10. Determining the Molecular Weight of a Gas: *Flick Your Bic* **17**
11. The Effect of Pressure on Boiling Point **19**

Solubility and Solutions

12. Precipitate Formation: White **23**
13. Precipitate Formation: Black and White **24**
14. Precipitate Formation: Blue **25**
15. Effect of Temperature on Solubility **26**
16. Negative Coefficient of Solubility: Calcium Acetate **27**
17. Supersaturation and Crystallization **28**
18. The Silicate Garden **30**
19. The Effect of Temperature on a Hydrate: Pink to Blue **31**
20. Cobalt Complexes: Changing Coordination Numbers **32**
21. Polar Properties and Solubility **34**

Acids and Bases

22. Acid–Base Indicators **37**
23. Acid–Base Indicators: Universal Indicator **38**
24. Acid–Base Indicators and pH **39**
25. Acid–Base Indicators: A *Voice-Activated* Chemical Reaction **40**

Energy Changes

26. Endothermic Reaction: Ammonium Nitrate **43**
27. Endothermic Reaction: Two Solids **44**

28. Endothermic Reaction: Thionyl Chloride and Cobalt Sulfate **45**
29. Exothermic Reaction: Calcium Chloride **46**
30. Exothermic Reaction: Sodium Sulfite and Bleach **47**

Equilibrium

31. Equilibrium and LeChatelier's Principle **51**
32. Effect of Temperature Change on Equilibrium: Cobalt Complex **53**
33. Effects of Concentration and Temperature on Equilibrium: Copper Complex **54**
34. Effect of Concentration on Equilibrium: Cobalt Complex **56**
35. Equilibrium: The Chromate–Dichromate System **58**
36. Equilibrium in the Gas Phase **60**
37. Equilibrium: Temperature and the Ammonium Hydroxide–Ammonia System **62**
38. Equilibrium: Effect of Temperature **63**
39. Effect of Pressure on Equilibrium **64**
40. Effect of Hydrolysis on Equilibrium **65**
41. Solubility Product: Effect of Concentration **66**

Kinetics

42. Carbon as a Catalyst **69**
43. Medicine Cabinet Kinetics: How Fast Is the Fizz? **70**
44. Catalytic Decomposition of Hydrogen Peroxide: Foam Production **71**
45. A Catalyst in Action **72**
46. Autocatalysis **73**
47. The Starch–Iodine Clock Reaction **75**
48. The *Old Nassau* Clock Reaction **77**
49. Disappearing Orange Reaction: Now You See It, Now You Don't! **78**
50. A Traffic Light Reaction **79**
51. The Quick Gold Reaction **80**
52. An Oscillating Reaction: Clear–Brown **81**
53. An Oscillating Reaction: Yellow–Blue **82**
54. An Oscillating Reaction: Red–Blue **83**

Oxidation–Reduction

55. Catalytic Oxidation of Ammonia **87**
56. Oxidation of Glycerin by Permanganate **88**
57. Oxidation–Reduction: Iron **89**
58. The Silver Mirror Reaction **91**
59. The Blue Bottle Reaction **93**
60. Oxidation of Zinc: Fire and Smoke **95**
61. The Mercury Beating Heart **96**
62. Oxidation States of Manganese: Quick Mn^{6+} **98**
63. Oxidation States of Manganese: Mn^{7+}, Mn^{6+}, Mn^{4+}, and Mn^{2+} **99**
64. The *Prussian Blue* Reaction **101**
65. The Activity Series for Some Metals **103**
66. Copper into Gold: The Alchemist's Dream! **104**

67. Oxidation of Sodium **105**
68. Oxidation of Alcohol by Mn_2O_7 **106**
69. Oxidation States of Vanadium: Reduction of V^{5+} to V^{2+} **108**
70. Oxidation States of Vanadium: Reoxidation of V^{2+} to V^{5+} **110**
71. Oxidation States of Chromium **112**
72. Photoreduction: The *Blueprint* Reaction **113**

Electrochemistry

73. Making a Simple Battery: The *Gerber* Cell **117**
74. Displacement of Tin by Zinc **119**

Other Chemical Reactions

75. Double Displacement: Reaction Between Two Solids **123**
76. Dehydration of Sucrose **124**
77. Dehydration of *p*-Nitroaniline: Snake and Puff! **125**
78. Synthesis of Nylon **126**
79. Synthesis of Rayon **128**
80. Synthetic Rubber **130**
81. A Chemical Sunset **132**
82. Production of Sterno: A Gel **134**
83. Production of a Foam **135**
84. Another Foam **136**
85. Surface Tension of Water: The Magic Touch **137**
86. Chemiluminescence: The Firefly Reaction **138**
87. Chemiluminescence: Two Methods **140**
88. The Decomposition of Ammonium Dichromate: The *Volcano Reaction* **142**
89. Color Changes in Fe(II) and Fe(III) Solutions **143**
90. The Mello-Yello Reaction **145**
91. *Magic* Writing Reactions **147**
92. Patriotic Colors: Red, White, and Blue **148**
93. Snakes Alive! **149**
94. A Chemical Pop Gun **150**
95. Colored Flames **151**
96. Metal Trees **153**
97a. The Common Ion Effect: First Demonstration **154**
97b. The Common Ion Effect: Second Demonstration **155**
98. The Common Ion Effect: Lead Chromate **157**
99. The Common Ion Effect: Ammonium Hydroxide and Ammonium Acetate **159**

Smoke, Fire, and Explosions

100. Instant Fire **163**
101. Production and Spontaneous Combustion of Acetylene **164**
102. Oxidation of Phosphorus: *Barking Dogs* **166**
103. An Explosion: The Rapid Oxidation of Phosphorus **168**
104. A Puff of Smoke **169**
105. The Methanol Cannon **170**

106. Smoke Rings **171**
107. A Simple Explosive: Nitrogen Triiodide **173**
108. An Explosive Combination of Hydrogen and Oxygen **175**

Appendixes

Appendix 1. The Periodic Table: Electronegativity, Atomic Diameters, and Ionization Energy **179**
Appendix 2. Properties and Preparation of Laboratory Acids and Bases **180**
Appendix 3. Equipment and Reagent List **181**
Appendix 4. Safety and Disposal **185**

Index **189**

Preface

This book contains more than 100 demonstrations suitable for use with an introductory chemistry program. These examples were selected because they are simple, safe, effective, and enjoyable. Further, they can be used to introduce many of the major concepts in chemistry.

Our purpose in presenting these demonstrations is not only to provide the chemistry teacher with a sourcebook of ideas, but also to promote chemical demonstrations as a teaching technique to be used with the blackboard, textbook, and laboratory. Performing effective demonstrations in class can make chemistry more understandable and can be fun for student and teacher.

We suggest that the teacher keep the following in mind regarding demonstrations:

1. Demonstrations should not be thought of as a replacement for the laboratory. Nothing can substitute for the hands-on experience provided by the laboratory. Rather, demonstrations should be thought of as an extension of the laboratory—another opportunity for students to become more astute observers.
2. Demonstrations should actively involve students. Chemistry is not a *spectator sport!* Although the teacher actually performs the demonstrations, students can be involved as assistants whenever possible.
3. Demonstrations should be simple and easy to understand. Quite often we overlook very effective demonstrations because they seem so simple. The demonstrations in this book do not require exotic chemicals or elaborate equipment, nor do they introduce concepts outside the scope of the general chemistry program.
4. Demonstrations should catch and hold student interest. They should be short and catchy. To achieve the latter goal, the demonstrations in this book involve color changes, gas evolution, precipitate formation, smoke, fire, and other obvious chemical changes. They are intentionally designed to be enjoyed by the student!
5. Demonstrations should work. They must always be rehearsed before class. Even if a demonstration has been performed dozens of times, it should still be checked to ensure that all solutions are still good. Remember, to experiment means to TRY, but to demonstrate means to SHOW.

These demonstrations are presented in a simple format for quick reference: You are told what the demonstration shows, how to do it, what the reactions are, and how to prepare the solution. Additionally, we offer some teaching tips, including notes and questions suitable for use with the demonstrations.

You are urged to take necessary precautions to ensure the safety of yourself and your students. Although the demonstrations in this book are generally safe, we have presented in a special section a few demonstrations that require additional safety precautions. Although these produce dramatic effects, special care must be taken in preparing solutions and doing the demonstrations. Student involvement in the demonstrations in this section should be kept to a minimum. Some reactions, such as the thermite reaction, are considered dangerous and are not included in this book.

You are encouraged to modify these demonstrations to fit your own need. Add to them and begin your own collection. Perform them for your classes and, if the opportunity arises, put on a show for parents or a workshop for other teachers.

Proper acknowledgment of the originator of most demonstrations is difficult because the demonstrations have been around for a long time and have undergone many modifications—including ours. However, all of the efforts of chemists, past and present, who have tried to make chemistry more interesting and understandable by developing demonstrations are gratefully acknowledged. We also express our appreciation to the many chemistry teachers and students around the country who have used our demonstrations and offered suggestions for improving them.

Special thanks are due two individuals: Julie Fisher Ealy, who edited our manuscripts; and David Daniel, Manager of the Office of High School Chemistry of the American Chemistry Society. This office, through its Expert Demonstrator Training Activity workshops, has kindled a nationwide interest in chemical demonstrations. We also thank Bob Johnson, Acquisitions Editor for ACS, and Leeallyn Clapp, David Brooks, and Christie Borgford for reviewing the manuscript for revision.

Lee R. Summerlin
The University of Alabama at Birmingham
Birmingham, AL 35294

James L. Ealy, Jr.
The Hill School
Pottstown, PA 19464

Properties of Atoms

. Electronegativity, Atomic Diameter, and Ionization Energy

. model is prepared to produce a three-dimensional representation of the trends in electronegativity, tomic diameter, and ionization energy shown by the periodic arrangement of selected atoms.

rocedure

1. Prepare a board as shown in Figure 1. Cut out a portion of a periodic chart (or draw one) and paste it to the board.
2. With a punch drill or some sharp pointed instrument, make a hole in the center of each square to represent an atom on the chart.
3. Cut toothpicks in various lengths to represent the magnitude of the property to be illustrated.

   Electronegativity: 1 cm = 1 electronegativity unit
   Atomic Diameter: 1 cm = 1 Ångstrom
   Ionization Energy: 1 cm = 6 eV

   Refer to the periodic chart in Appendix 1 for correct values. Be sure to add 0.5 cm for the depth of the hole.
4. Place the toothpicks in the appropriate positions in the chart.

   Figure 1 shows toothpicks cut to represent the electronegativity of one period and one group of elements.

Teaching Tips

NOTES

1. Colored, round toothpicks work best. Use a different color for each property.
2. You can glue the toothpicks in place for a permanent classroom display.
3. Transitional elements are omitted because properties show little change from element to element.

QUESTIONS FOR STUDENTS

1. What is the general trend in electronegativity across a row of elements? How can you account for this?
2. What is the general trend in electronegativity down a column of elements? How can you account for this?
3. According to the model, which two elements should form the strongest ionic bond?

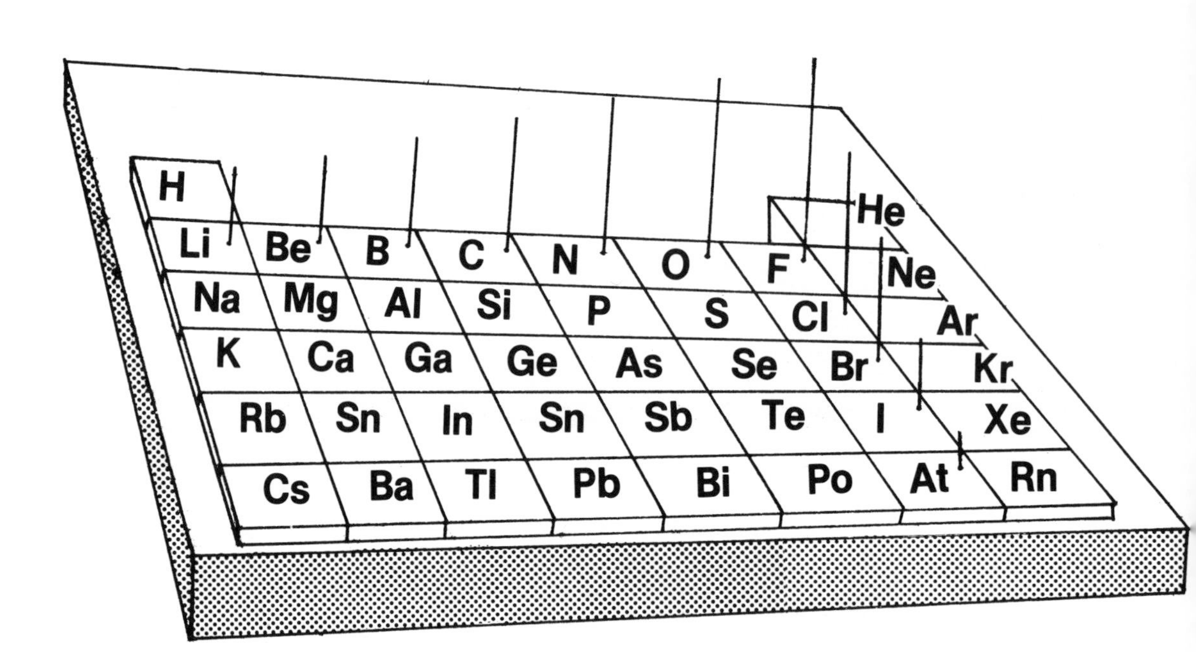

Figure 1. Board for model for properties of atoms.

Gases

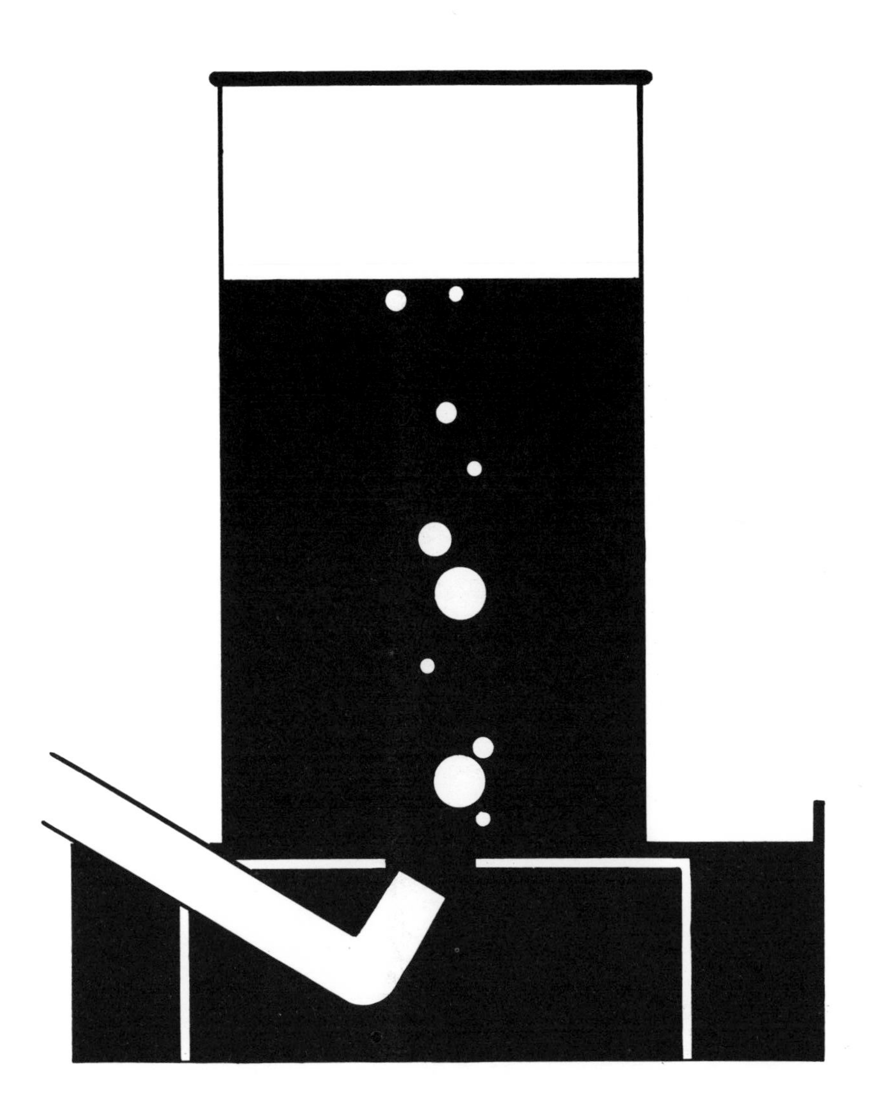

2. Gas Densities

Large bubbles are formed by bubbling laboratory gas through a soap solution. As the bubbles rise in the air they are *ignited* with a candle attached to a meter stick. Spectacular effect!

Procedure

1. Either obtain a small bottle of bubble solution at the toy store or prepare a detergent solution. Place the solution in a large beaker.
2. Carefully blow bubbles in the solution by using a hose attached to the laboratory methane gas outlet. A small funnel or thistle tube works well as a pipe.
3. Have an assistant standing by with a birthday candle taped to the end of a long stick.
4. As the bubbles rise in the air, ignite them with the candle.

Teaching Tips

NOTES

1. This demonstration illustrates that methane gas has a density less than that of air.
2. A little practice is needed to get good bubbles. As the bubbles form, gently shake them loose from the funnel and they will float upward.
3. Add a small amount of glycerin to produce a better bubble solution.

QUESTIONS FOR STUDENTS

1. This demonstration works well with methane gas. Would it also work with propane? Butane?
2. Would it work with hydrogen? (Try it!)

3. Properties of Gases: Pressure and "Suction"

A deflated balloon is placed between the mouths of two cups. As the balloon is inflated, the cups adhere to the balloon and will not fall off as the balloon is passed around the class.

Procedure

1. Obtain several balloons that will easily inflate to 6–8 in.
2. Hold two cups with the deflated balloon between them (you may want a student to assist you).
3. Blow into the balloon and inflate it with one breath.
4. Hold up the balloon with the attached cups and ask for an explanation.

Reaction

As the balloon is inflated, air pressure forces the sides of the balloon against the cups with such force that the cups adhere to the balloon. This sequence creates a "suction" effect on the cups.

Teaching Tips

NOTES

1. Cheap porcelain cups work well. If you are short on confidence, paper and styrofoam cups work just as well!
2. Try using containers of various size and shape.

QUESTIONS FOR STUDENTS

1. Can you give other examples of this phenomenon? (e.g., pipeting a solution or drinking through a straw).
2. Explain this phenomenon on a molecular basis.

4. Temperature and Pressure Relationships

A large flask containing an inflated balloon is shown to the class.

Procedure

1. Select an Erlenmeyer flask with a narrow mouth. Place about 5 mL of water in the flask.
2. Heat the flask until the water boils down to a volume of about 1 mL.
3. Remove the flask from the heat, hold it with a towel, and immediately place the open end of a colored balloon over the mouth of the flask.
4. Observe the effect as the flask cools—the balloon will be sucked into the flask.
5. To remove the balloon, heat the flask.

Reaction

As the flask cools, the pressure inside the flask decreases. Because the pressure outside the flask is greater, the balloon is *sucked* (pushed) into the flask.

Teaching Tips

NOTES

1. You will get better results if you partially inflate the balloon before attaching it to the hot flask.
2. Use this example as an inquiry demonstration. Show the flask to your students and ask for possible explanations. See if anyone notices the small amount of water in the flask.

QUESTIONS FOR STUDENTS

1. Why is it necessary to boil the water?
2. Why is the balloon sucked into the flask?
3. Would another liquid work just as well?
4. How can you remove the balloon (without breaking the flask)?

5. Solubility of a Gas: The Ammonia Fountain

The set-up shown in Figure 2 is displayed. The flask is filled with ammonia gas. The beaker is filled with water. When a small amount of water is squirted into the flask from the dropper, water rises in the glass tube and is sprayed into the flask as a pink fountain. The fountain continues until the flask is almost filled.

Procedure

1. Set up the apparatus as shown in Figure 2. Notice that the tube in the flask has a fine tip. The flask must be DRY.
2. Fill the beaker three-fourths with water and add a few drops of phenolphthalein indicator.
3. Fill the dropper with water.
4. Fill the flask with ammonia by the following process:

   a. Place a spoonful of ammonium chloride and a spoonful of sodium hydroxide in a large DRY test tube.
   b. Gently heat the tube IN A HOOD or well-ventilated area and direct the ammonia produced into the dry flask. Be sure to dispel all air from the flask and collect a full flask of ammonia gas.

5. Place a pinch-clamp on the tube inserted in the stopper and stopper the flask.
6. Place the flask upside-down in the ring on a ringstand.
7. Place the tube extending from the flask in the beaker. The end of the tube should be just above the bottom of the beaker.
8. When you are ready to begin the demonstration, remove the pinch-clamp.
9. Begin the reaction by squirting water from the dropper into the flask.
10. A fountain will result as the water from the beaker is sprayed into the flask.

Reactions

1. Production of ammonia:

$$NH_4^+(aq) + \text{excess } OH^-(aq) \rightleftharpoons NH_3(g) + H_2O(\ell)$$

2. The solubility of ammonia gas in water is so great that most of the ammonia immediately dissolves in the water from the dropper. This produces a partial vacuum in the flask. Because of the difference in pressure, water from the beaker rises and enters the flask.
3. The basic ammonium hydroxide produced reacts with the phenolphthalein indicator and turns the solution red.

Teaching Tips

NOTES

1. You can also produce ammonia gas in the flask by adding a small amount of concentrated ammonium hydroxide and gently warming the flask to produce ammonia.
2. If you place about 20 mL HCl in the beaker and add litmus rather than phenolphthalein as an indicator, the red solution will become blue as it enters the flask.
3. HCl is also very soluble in water. You can make HCl by the following process: Gently heat a test tube containing sodium chloride and concentrated sulfuric acid,

and direct the HCl produced into a dry flask. If you use litmus indicator, place a small amount of ammonium hydroxide in the beaker. If you use methyl violet indicator, you will produce yellow, green, blue, and violet solutions. Use only water and indicator in the flask.

4. Experiment with other indicators. Methyl orange works well.
5. If the demonstration doesn't work, it is probably because the flask was not dry or the flask was not completely filled with ammonia.

QUESTIONS FOR STUDENTS

1. Why will this reaction not work if the flask is wet?
2. Why is ammonia not collected by displacing water, as one would collect oxygen?
3. What other gases show great solubility in water?
4. Explain the reaction used to produce ammonia gas.

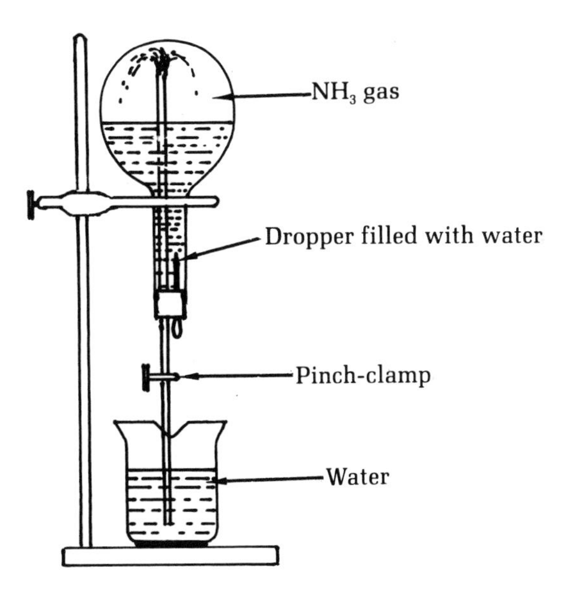

Figure 2. Set-up for the ammonia fountain.

6. Preparation of Oxygen Gas from Laundry Bleach

Oxygen gas is prepared by catalytic decomposition of laundry bleach. The gas is collected by displacement of water.

Procedure

1. Set up a gas-collecting apparatus. A large filtering flask with a hose connected to the side arm and leading into a shallow dish filled with water works well. Collect the gas by displacing water from filled test tubes.
2. Place 100 mL fresh laundry bleach in the filtering flask.
3. Add about 5 g (the amount is not critical) of cobalt(II) chloride to the flask.
4. Stopper the flask quickly and swirl gently to mix contents.
5. Oxygen gas will be produced. Before collecting the gas, displace all the air in the flask and test tubes.
6. Ignite a glowing splint to show the presence of oxygen gas.
7. Heat steel wool and place it in one of the test tubes.

Reaction

$$\underset{\text{(hypochlorite)}}{2ClO^-(aq)} \xrightarrow{\text{catalyst}} O_2(g) + 2Cl^-(aq)$$

Teaching Tips

NOTES

1. Laundry bleach contains sodium hypochlorite as the active ingredient.
2. The black precipitate is probably an unstable oxide, such as Co_2O_3, that decomposes to form oxygen, then recombines with the hypochlorite ion.
3. Because the rate of this reaction varies with temperature, it can be used to illustrate chemical kinetics:

   a. Add 3 mL of 0.2 M cobalt nitrate to 15 mL of bleach.
   b. Vary the temperature and compare the rates of reaction.

QUESTIONS FOR STUDENTS

1. Record your observations.
2. Write the equation for the reaction.
3. Why did we collect the gas by displacement of water?
4. Why did the splint burst into flame?
5. Why might this reaction rate vary with temperature?

7. Preparation of Chlorine Gas from Laundry Bleach

Chlorine gas and chlorine water are prepared by reacting laundry bleach with hydrochloric acid. The bleaching property of chlorine is shown.

Procedure

1. Prepare a gas-generating system by connecting a rubber tube to the side arm of a large filtering flask.
2. Place 30 mL of laundry bleach in the flask.
3. Add 5 mL of HCl. Stopper the flask and swirl it quickly.
4. Collect chlorine gas by upward displacement of air in several test tubes.
5. Test for the presence of chlorine by the usual methods (e.g., ability to bleach non-colorfast fabric or dyes).

Reaction

$$ClO^-(aq) + Cl^-(aq) + 2H^+(aq) \rightarrow Cl_2(g) + H_2O(\ell)$$

(hypochlorite)

Solution

The HCl concentration is 1.0 M (*see* Appendix 2). To generate chlorine faster, use more concentrated acid.

Teaching Tips

NOTES

1. Use a HOOD or well-ventilated area for this demonstration. AVOID DIRECTLY BREATHING THE CHLORINE.
2. This method is an easy way to prepare chlorine water.
3. Because bleach is prepared commercially by bubbling chlorine gas through sodium hydroxide, this demonstration is essentially the reverse of this reaction:

$$Cl_2(g) + 2OH^-(aq) \rightarrow ClO^-(aq) + Cl^-(aq) + H_2O(\ell)$$

4. Place a small piece of cloth soaked in turpentine in one of the tubes. CAUTION!

QUESTIONS FOR STUDENTS

1. Why was this gas not collected by the displacement of water?
2. What property of the gas allows us to collect it by the upward displacement of air?
3. Write the reaction for the preparation of chlorine gas.
4. How does bleach do what it does? (Nascent oxygen!)
5. What happened with the turpentine? Was this a *spontaneous* reaction?

8. Diffusion of Gases

A plug of cotton dipped in HCl is inserted in the open end of a graduated cylinder. The time required for a color change to occur in a strip of pH paper at the other end of the cylinder is noted. The procedure is repeated with a cotton plug dipped in NH_4OH. From these observations, Graham's law of diffusion is checked.

Procedure

1. Clean and DRY two 100-mL graduated cylinders.
2. With a glass rod, insert a moist piece of blue litmus paper in one cylinder. Push the paper to the bottom of the cylinder and see that it sticks to the bottom.
3. Place the cylinder on its side, making sure that it is level.
4. Dip a small piece of cotton in concentrated HCl (CAREFUL!) and place it just inside the opening of the cylinder.
5. Immediately place a piece of plastic wrap tightly over the opening of the cylinder.
6. As soon as the cotton plug is inserted, have a student assistant begin timing. Record the time required for the HCl gas to travel the length of the cylinder and cause the blue litmus paper to turn red.
7. Repeat the demonstration with red litmus paper and a cotton plug dipped in ammonium hydroxide in the other graduated cylinder.

Calculations

1. The data collected from this demonstration will be used to check Graham's law of diffusion of gases: The rate of diffusion of a gas is inversely proportional to the square root of its molecular mass.
2. Determine the rate: Divide the distance the gas traveled (length of the cylinder, in centimeters) by time (seconds).
3. Determine the ratio of the experimental diffusion rate: Divide the rate of diffusion of hydrogen chloride gas by the rate of diffusion of ammonia gas.
4. Check your answer against the theoretical ratio:

$$\frac{R_{NH_3}}{R_{HCl}} = \sqrt{\frac{M_{HCl}}{M_{NH_3}}} = 1.46$$

Teaching Tips

NOTES

1. Graduated cylinders are used because they are readily available and provide a short distance for the gas to travel.
2. An alternate method is to use a long piece of glass tubing. Simultaneously insert a cotton plug soaked in HCl in one end of the tube and a cotton plug soaked in ammonium hydroxide in the other end. Note the appearance of a white ring of ammonium chloride in the tube. Measure the distance from each end of the tube to this ring, and use this data to calculate the rate. (The white ring is often difficult to detect!)

QUESTIONS FOR STUDENTS

1. Why did the ratio calculated from the data collected in this demonstration differ from the theoretical ratio? (This is a good place to discuss experimental error.)
2. What is the relationship between the mass of gas and its rate of diffusion?
3. Why is it necessary to use DRY cylinders for the demonstration?
4. How would the calculated ratio be effected if the cylinders were not level during the demonstration?

9. Production of a Gas: Acetylene

A small amount of water and calcium carbide is placed inside a rubber balloon. The balloon is tied shut and soon begins to expand as a result of the production of acetylene gas. The expanded balloon is tied to a meter stick and held near a burner flame. (CAUTION!) A loud explosion results!

Procedure

1. With a dropper, squirt 2–3 mL of water into a balloon.
2. Push a piece of calcium carbide, CaC_2, through the neck of the balloon. Pinch the balloon to hold it in place while the neck of the balloon is securely tied.
3. Release the balloon and watch as it expands.
4. When the reaction and balloon expansion cease, tape the balloon to the end of a meter stick and hold it near a burner flame.

Reactions

1. Production of acetylene:

$$CaC_2\,(s) + 2H_2O(\ell) \rightarrow Ca(OH)_2(s) + C_2H_2(g)$$

2. Explosion of acetylene

$$2C_2H_2(g) + 5O_2(g) \rightarrow 4CO_2(g) + 2H_2O(g) + \text{heat}$$

Teaching Tips

NOTES

1. Practice this demonstration to get just the right amounts of water and calcium carbide for maximum effect.
2. This method is safer than the *exploding can* method of producing acetylene.
3. "Twist-ties" work well to separate water and CaC_2 before mixing.

QUESTIONS FOR STUDENTS

1. What suggestions can you give for storing calcium carbide?
2. Can you draw the structure of acetylene? ($HC \equiv CH$)
3. What are some commercial uses for acetylene?

0. Determining the Molecular Weight of a Gas: *Flick Your Bic*

large graduated cylinder is filled with water and placed, inverted, in a water-filled trough. A small piece of ubber tubing from a pocket lighter is placed inside the cylinder. When the release button on the lighter pressed, butane is released, displacing the water in the cylinder. By measuring the volume of water isplaced and the weight of the gas, the molecular mass of butane is calculated.

rocedure

1. Remove the striking mechanism (flint, wheel, and spring) from a NEW disposable pocket lighter.
2. Weigh the lighter on a balance. If you plan to use a small cylinder and displace a small volume of water, use an analytical balance. If you use a large (500 mL) cylinder, you can use a triple-beam balance. Record the weight.
3. Attach a small rubber tube to the gas nozzle of the lighter.
4. Fill the largest graduated cylinder you have with water, invert it, and place it inside a trough half-filled with water. Be sure that the cylinder contains no gas bubbles.
5. Place the end of the rubber tube under and inside the cylinder and press the release button on the lighter.
6. Collect enough gas to displace 300–400 mL of water, or the largest volume possible.
7. Carefully measure the water remaining in the cylinder to determine the gas volume.
8. Remove the tube from the lighter and reweigh the lighter.
9. From these data, calculate the molecular mass of butane. (Remember to subtract the vapor pressure of water—see box—and to record the temperature and pressure.)

alculations

1. PV equals nRT, where n is grams per molar mass. Thus, PV equals (grams/molar mass) RT.
2. Molar mass equals (grams) RT/PV, where R is the gas constant, 62,400 mL Torr$\cdot$mol$^{-1}\cdot$K^{-1}; T is the temperature in Kelvins (273 K = 0 °C); P is the pressure in Torr corrected for the vapor pressure of water; and V is the volume of gas in milliliters.

eaching Tips

NOTES

1. The molecular mass of butane is 58.
2. Now is a good time to discuss experimental error.
3. If the tube on the lighter leaks, your results will be off. Try placing the lighter directly beneath the water-filled cylinder and releasing the gas.
4. A hair dryer is handy to dry the lighter prior to final weighing.

QUESTIONS FOR STUDENTS

1. How can you explain the fact that butane is a liquid in the lighter, but a gas when it is collected?
2. If you only have a triple-beam balance, why is it necessary to collect a large volume of gas?

3. Why is it necessary to subtract the vapor pressure of water?
4. How does the vapor pressure of water vary with temperature?

Vapor Pressure of Water

Temperature (°C)	Pressure (Torr)
15	12.8
16	13.6
17	14.5
18	15.5
19	16.5
20	17.5
21	18.6
22	19.8
23	21.0
24	22.4
25	23.7
26	25.2
27	26.7
28	28.3
29	30.0
30	31.8

11. The Effect of Pressure on Boiling Point

A syringe is half-filled with water at a temperature below the boiling point. The end of the syringe is closed. When the plunger is pulled, the pressure is decreased, and the water boils.

Procedure

1. Heat water in a beaker to about 80 °C, well below boiling.
2. Using a large glass or plastic syringe with a short piece of rubber tubing attached, draw about 40–50 mL of hot water into the syringe.
3. Holding the syringe upright, push in the plunger to dispel any air in the syringe.
4. Clamp the tube tightly with a screw-type buret clamp.
5. Holding the syringe with the plunger up, slowly but forcefully pull on the plunger.
6. As the plunger is raised, the pressure on the hot water is reduced and the water boils.

Reaction

The boiling point of a liquid is the temperature at which vapor pressure of the liquid equals atmospheric pressure. For water, this temperature is 100 °C. If the atmospheric pressure is reduced, a lower temperature will provide a vapor pressure equal to this pressure and the water will boil at the lower temperature.

Teaching Tips

QUESTIONS FOR STUDENTS

1. What is the relationship between temperature, pressure, and boiling?
2. Why do bubbles appear at or near the boiling point?
3. Would this demonstration work with another liquid, such as alcohol? (If you try this method, used a *hot plate* to heat the alcohol.)

Solubility and Solutions

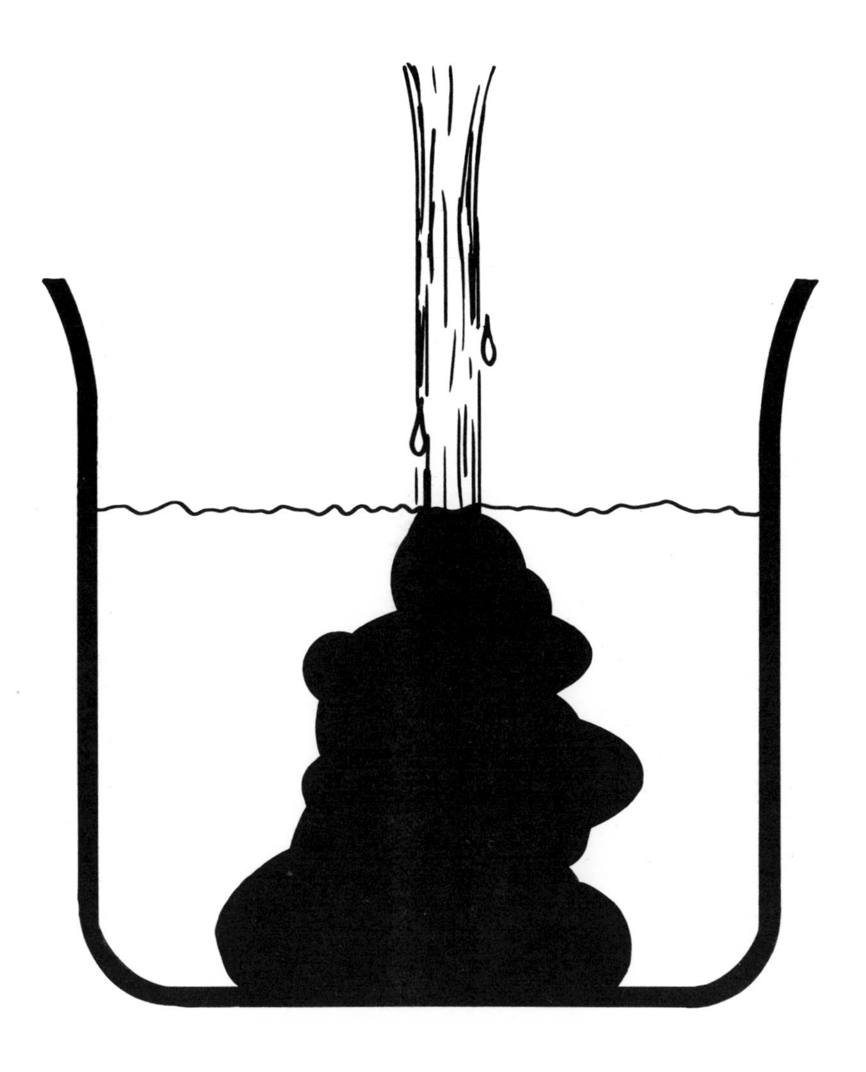

12. Precipitate Formation: White

Two colorless solutions are mixed. In 1–2 min, a copious, flaky white precipitate forms.

Procedure

1. Place 180 mL of solution B in a large beaker.
2. Add 40 mL of solution A. DO NOT STIR.
3. After a few minutes a heavy precipitate will form.
4. Place the beaker on the demonstration desk and note the continued precipitation over the next 5–10 min.

Reaction

$$2K(SbO)C_4H_4O_6(aq) + BaCl_2(aq) \rightarrow \underset{\text{(precipitate)}}{Ba[(SbO)C_4H_4O_6]_2} + 2KCl(aq)$$

Solutions

1. Solution *A*, barium chloride: Dissolve 10 g of $BaCl_2$ in 40 mL of distilled water.
2. Solution *B*, antimony potassium tartrate: Dissolve 27 g of $K(SbO)C_4H_4O_6$ in 180 mL of distilled water.
3. It may be necessary to gently heat both solutions.

Teaching Tips

NOTES

1. Antimony potassium tartrate is also called potassium antimony tartrate.
2. If desired, the precipitate can be collected, washed with alcohol, and dried.

QUESTIONS FOR STUDENTS

1. Write the chemical equation for this reaction.
2. Even though the formulas are complex, what type of reaction does this demonstration show?
3. What causes a precipitate to form when the two solutions are mixed?

13. Precipitate Formation: Black and White

Two clear solutions are placed in separate beakers labeled 1 and 2. When the contents of beaker 1 are poured into beaker 2, a white precipitate immediately forms. However, when the process is repeated and the contents of beaker 2 are poured into beaker 1, a black precipitate is formed. Two different results are produced by mixing the same solutions!

Procedure

1. Place 10 mL of $SnCl_2$ solution in beaker 1.
2. Place 90 mL of $HgCl_2$ solution in beaker 2.
3. Rapidly pour the contents of beaker 2 into beaker 1.
4. Note the formation of a white precipitate.
5. Place 10 mL of $HgCl_2$ solution in beaker 1.
6. Place 90 mL of $SnCl_2$ solution in beaker 2.
7. Rapidly pour the contents of beaker 2 into beaker 1.
8. Note the formation of a black precipitate.

Reactions

The first reaction is

$$2Hg^{2+}(aq) + Sn^{2+}(aq) + 2Cl^{-}(aq) \rightarrow \underset{\text{(white precipitate)}}{Hg_2Cl_2(s)} + Sn^{4+}(aq)$$

The second reaction is

$$HgCl_2(s) + Sn^{2+}(aq) \rightarrow \underset{\text{(black precipitate)}}{Hg(s)} + Sn^{4+}(aq) + 2Cl^{-}(aq)$$

Solutions

1. The $SnCl_2$ solution is 0.1 M: 19 g of $SnCl_2$ per liter of water.
2. The $HgCl_2$ solution is 0.1 M: 27.2 g of $HgCl_2$ per liter of water.
3. Add a drop or two of concentrated HCl to each solution to prevent the formation of metal–hydroxy complexes.

Teaching Tips

NOTES

1. The $SnCl_2$ solution must be made fresh prior to the demonstration.
2. Remember, the larger volume is always poured into the smaller volume.

QUESTIONS FOR STUDENTS

1. Write the chemical equations for the two reactions.
2. What are the formula and name of the white precipitate?
3. What are the formula and name of the black precipitate?
4. Why do they form differently?
5. Do the different volumes of solutions have an effect on the reaction?

4. Precipitate Formation: Blue

small amount of wine-colored solution is added to a clear solution. A blue precipitate forms.

rocedure

1. Place ~ 200 mL of clear limewater in a beaker.
2. Add a few drops of cobalt chloride solution from a dropper.
3. Note the immediate formation of a blue precipitate.

eaction

$$CoCl_2(aq) + Ca(OH)_2(aq) \rightarrow Ca^{2+}(aq) + 2Cl^-(aq) + \underset{\text{(blue precipitate)}}{Co(OH)_2(s)}$$

olutions

1. The limewater, $Ca(OH)_2$, is a saturated solution. Let it stand overnight and pour off the clear supernate.
2. The concentration of cobalt chloride, $CoCl_2$, is not critical; try various amounts.

eaching Tips

NOTES

1. This method is a good demonstration to show students that all precipitates are not white!
2. This demonstration projects well; use petri dishes on an overhead projector.

QUESTIONS FOR STUDENTS

1. Write the equation for the reaction.
2. What do you know about the general solubility of hydroxides?
3. What is the blue precipitate?
4. Could you do this demonstration by adding cobalt ion to another hydroxide such as NaOH? Try it!

15. Effect of Temperature on Solubility

A large stoppered test tube containing a clear solution is placed in a mixture of salt and ice, and pink crystals separate out. The tube is then heated and the solution becomes clear again. Upon further heating a white precipitate forms!

Procedure

1. Prepare a mixture of salt and ice to provide a very low-temperature solution.
2. Place the test tube in the mixture. Note formation of pink crystals.
3. Heat the tube to 27 °C. Note that the solution becomes clear.
4. Continue to heat the tube above 27 °C. Note the formation of a white precipitate.

Reactions

1. The pink crystals are $MnSO_4 \cdot 7H_2O$.
2. The white precipitate is $MnSO_4 \cdot H_2O$.

$$\underset{\text{(pink precipitate)}}{MnSO_4 \cdot 7H_2O(s)} + \text{heat} \rightleftharpoons \underset{\text{(clear solution)}}{Mn^{2+}(aq) + SO_4^{2-}(aq)} + \text{heat} \rightleftharpoons \underset{\text{(white precipitate)}}{MnSO_4 \cdot H_2O(s)}$$

Solutions

Prepare the solution in the test tube as follows:

1. Add a few drops of concentrated sulfuric acid to 100 mL of water at 27 °C.
2. Add EITHER 70 g of $MnSO_4$ or 130 g of $MnSO_4 \cdot 7H_2O$.
3. Decant the clear solution into a test tube and seal the tube.

Teaching Tips

NOTES

1. This demonstration effectively shows that solubility generally increases with temperature, but that this rule does have exceptions.
2. Have your students determine the temperature of the salt and ice mixture.

QUESTIONS FOR STUDENTS

1. Obviously this reaction is temperature dependent. Why?
2. Do the precipitates contain water of hydration?
3. What happens at a higher temperature?
4. Write the formulas for the two precipitates.
5. Can you think of other compounds that might behave the same way?

16. Negative Coefficient of Solubility: Calcium Acetate

A clear solution is heated and a precipitate forms. When the solution is cooled, the precipitate dissolves.

Procedure

1. Place 150 mL of calcium acetate solution in an Erlenmeyer flask and heat the flask. Record the temperature.
2. Note that a precipitate begins to form at ∼ 80 °C.
3. Remove the flask from the heat and cool it by placing it in a stream of cold water. Record the temperature.
4. Notice that the solid goes back into solution as the temperature decreases.

Solution

The calcium acetate solution is saturated (about 40 g/100 mL of water).

Teaching Tips

NOTES

1. This example is a good demonstration to show students that most rules in chemistry have exceptions.
2. The difference in solubility at the two temperature extremes is about 15 g/100 g water: at 0 °C, the solubility is 44–52 g/100 g water; at 100 °C, the solubility is 36–45 g/100 g water.

QUESTIONS FOR STUDENTS

1. What appears to happen?
2. Is this what you would normally expect? What usually happens?
3. Does water of hydration enter into the results?
4. Devise a model for the expected results.

17. Supersaturation and Crystallization

A clear solution is poured over a few crystals on a laboratory bench. As the solution is poured, a crystal matrix is formed, and a tall crystal column is produced.

Procedure

1. Clean off an area of the lab bench.
2. Place a few small crystals of sodium acetate on the clean area.
3. Slowly drip the solution over the crystals.
4. Crystallization will begin and a tall column of crystals can be formed.

Reactions

A supersaturated solution contains more than the normal saturation quantity of a solute. When a small crystal of solid solute is added, crystallization of the excess solute results, and an equilibrium appropriate to the lower temperature is restored.

$$NaCH_3COO\ (s) \rightleftharpoons Na^+\ (aq) + CH_3COO^-\ (aq)$$

Solutions

Prepare the sodium acetate solution as follows:

1. Place 50 g of sodium acetate trihydrate in a small flask.
2. Add 5 mL of water and slowly warm the flask.
3. Swirl the flask until the solid completely dissolves. If any solid remains on the neck or sides of the flask, wash it down with a small amount of water.
4. Remove the flask from the heat, cover it with aluminum foil, and allow it to cool at room temperature.

Alternate Method

1. Clean and dry a small Erlenmeyer flask and a stopper that fits.
2. Fill the flask with solid sodium acetate trihydrate.
3. Slowly heat the flask on a hot plate until it completely liquefies.
4. The solid should just melt, not boil!
5. Using a squirt bottle, carefully rinse the neck of the flask with a small amount of water.
6. Insert the stopper and allow the flask to cool at room temperature.
7. Remove the stopper when the flask has cooled, and carefully add only *one* crystal of solid sodium acetate trihydrate to the flask.
8. Observe! You can reheat the flask and use the same material over and over.

Teaching Tips

NOTES

1. This demonstration is a nice variation of the usual reaction in which a solution is *seeded* by adding a small crystal to a supersaturated solution.
2. Practice to get just the right *touch* necessary to form a tall column. The solid sodium acetate can be reused.

3. Sodium acetate is in the form of a trihydrate.
4. If you fill a clean buret with the saturated solution, a tall column of crystal will form as the solution drips from the buret tip onto a few crystals on the lab bench.

QUESTIONS FOR STUDENTS

1. **What has happened?**
2. **Why did the column build upward?**
3. **What reaction was taking place? Is this a chemical or a physical reaction?**
4. **Why was the flask wrapped in foil and slowly cooled?**

18. The Silicate Garden

A few small, colored crystals are added to a solution in a large jar or bowl. In a few seconds large, plant-like growths extend from the crystals.

Procedure

1. Select a wide-mouth container (a small fish bowl works nicely) and fill it to within 1 in. of the top with sodium silicate solution.
2. Drop in 3–4 small (matchhead size) crystals.
4. Notice the growth of the crystals within a few seconds.

Reactions

1. When metal salts are added to a silicate solution, insoluble silicates are formed.
2. When the salts are placed in the sodium silicate solution, a semipermeable membrane is formed around the salt. Because the concentration is greater inside the membrane, water enters the membrane to dilute the concentrated solution. This effect is called *osmosis.*

   Osmosis causes the membrane sack to break. It breaks upward because the pressure of water on the sides of the crystal is greater than on the top. This process is repeated as a new membrane forms, and an upward growth of the crystal garden results.

Solutions

1. The sodium silicate (also called *water glass*) is a diluted solution with a specific gravity of about 1.10. A dilution of 1 part sodium silicate to 4 parts water is a good approximation.
2. Use the following crystals to get a variety of colors in the *garden:* ferric chloride, brown; nickel nitrate, green; cupric chloride, bright green; uranyl nitrate, yellow; cobaltous chloride, dark blue; cobaltous nitrate, dark blue; manganous nitrate, white; and zinc sulfate, white.

Teaching Tips

NOTES

1. Covering the bottom of the container with a thin layer of sand will prevent the crystals from sticking to the bottom of the glass.
2. You can keep this as a permanent display. If the solution becomes cloudy after a few days, carefully remove it and replace it with water.
3. This method is an excellent way to demonstrate *osmosis.*

QUESTIONS FOR STUDENTS

1. Why do the crystal trees grow upward?
2. A large crystal appears to grow from a small crystal. Explain this.
3. How is *osmosis* involved in this phenomenon?
4. Is the growth continuous? If not, explain.
5. Would crystals of other chemicals produce such growths? (Try them!)

19. The Effect of Temperature on a Hydrate: Pink to Blue

A beaker containing a pink solution is heated on a hot plate. As the solution becomes warm, the color changes to blue. The beaker is removed from the heat and allowed to cool. The pink color returns.

Procedure

1. Place ~ 250 mL of 95% ethyl alcohol in a large beaker. Let it sit until it reaches room temperature.
2. Add ~ 1.5 g of cobalt(II) chloride and stir until it is dissolved. The solution should be pink.
3. Place the beaker on a hot plate and warm the solution. The solution will turn blue.
4. Remove the beaker from the hot plate. As it cools, the solution becomes pink again.

Reactions

1. Cobalt(II) chloride is in the form of a hydrate. When the solution is heated, the blue anhydrous form is produced.

$$\underset{\text{(pink)}}{Co(H_2O)_6^{2+}} + 4Cl^- \rightleftharpoons \underset{\text{(blue)}}{CoCl_4^{2-}} + 6H_2O(\ell)$$

2. When the solution is cooled, water in the solution recombines with the cobalt(II) chloride. The pink hydrate is again formed.

Solution

Use 95% ethyl alcohol, not pure ethyl alcohol. You may need to add water to 95% alcohol until you get a pink color.

Teaching Tips

NOTES

1. This reaction is essentially the same as that produced in "weather indicators." Strips of blue anhydrous cobalt(II) chloride paper turn pink when the humidity is high.
2. Demonstrations 20, 32, and 34 offer variations of this reaction.

QUESTIONS FOR STUDENTS

1. Why is 95% alcohol used in this demonstration?
2. Show the reaction between the pink and blue forms of this compound.
3. What part does heat play in this reaction?

20. Cobalt Complexes: Changing Coordination Numbers

A small amount of cobalt chloride solution is added to test tubes containing varying amounts of ethyl alcohol. The solutions in the tubes turn red, violet, and blue, color changes representing changes in the coordination number of cobalt chloride from 6 to 4.

Procedure

1. Fill a small test tube one-half with ethyl alcohol.
2. Fill a second tube one-fourth, and a third one-eighth with ethyl alcohol.
3. Add a dropper of cobalt chloride solution to each tube.
4. Note the change in color in each tube.

Reactions

1. The general reaction is:

$$\underset{\text{(pink)}}{Co(H_2O)_6^{2+}} + 4Cl^- \rightleftharpoons \underset{\text{(blue)}}{CoCl_4^{2-}} + 6H_2O(\ell)$$

2. As molecules of ethyl alcohol replace the water molecules in the coordination positions of the complex ion, the color of the solution changes. When one alcohol is attached, the color is red; when two are attached, the color is violet; when three are attached, the solution is a deep blue.
3. The coordination number of the cobalt(II) ion changes from 6 to 4.

Solutions

1. The ethyl alcohol solution is 100%.
2. The cobalt(II) chloride solution is 2 M: Dissolve 2.6 g of $CoCl_2$ in 10 mL of water.

Teaching Tips

NOTES

1. You can also demonstrate the color change from the hydrated to the dehydrated form with a *magic* writing reaction:

   a. Write a message on white paper with the pink hydrated solution.
   b. When the paper is dry, gently warm it.
   c. Dehydration of the salt will produce the blue dehydrated form, and the message will appear.

2. The blue [$CoCl_4^{2-}$] ion is tetrahedral. (See structure on next page.)

QUESTIONS FOR STUDENTS

1. What are *coordination numbers*?
2. How do the coordination numbers change with the addition of alcohol molecules?

3. How are complex ions with various coordination numbers characterized by colors?
4. Show the reaction for the formation of the blue anhydrous complex.

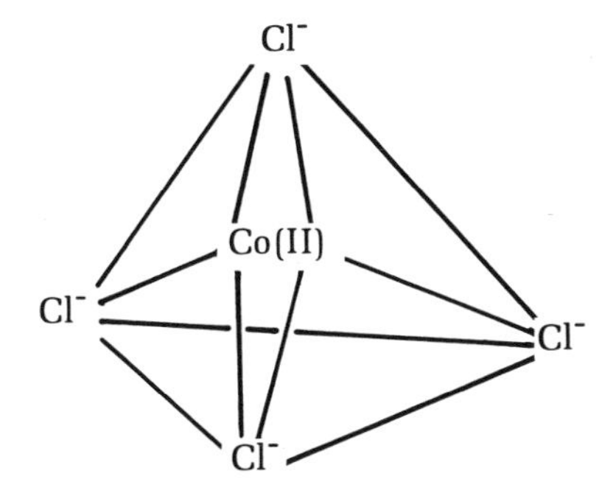

21. Polar Properties and Solubility

A large graduated cylinder is filled with, seemingly, a single liquid. A deflagrating spoon containing a few crystals of iodine is lowered into the top part of the liquid. The solution becomes dark in color. The spoon is lowered midway into the cylinder and no color results. The spoon is then lowered to the bottom of the cylinder and a violet color is produced, resulting in three distinct layers of liquids in the cylinder.

Procedure

1. Pour 150 mL of carbon tetrachloride into a large graduated cylinder.
2. Slowly add 150 mL of water; allow it to run down the side of the cylinder to avoid mixing.
3. Add 150 mL of petroleum ether; again, allow it to run down the side of the cylinder. Prepare this solution in advance, so that the cylinder appears to contain only one liquid.
4. Place a few crystals of iodine in a deflagrating spoon. Slowly lower the spoon into each liquid layer.

Reactions

1. Iodine, carbon tetrachloride, and ether are nonpolar substances and thus dissolve in each other. Therefore, the characteristic colors of iodine in solution are produced.
2. Water is a polar substance and does not readily dissolve a nonpolar substance, like iodine.

Solutions

Because many organic solvents have been removed from school laboratories, you may have to try other combinations. Carbon tetrachloride, methylene chloride, benzene, *p*-xylene, petroleum ether, and methyl alcohol work well.

Teaching Tips

NOTES

1. Take the usual precautions when handling organic solvents.
2. If you don't have a deflagrating spoon, drop a few large pieces of iodine into the cylinder.
3. This demonstration is an excellent way to introduce molecular polarity.

QUESTIONS FOR STUDENTS

1. What prevents the three liquids from mixing?
2. Why are the colors in the top and bottom layer different?
3. What is a *tincture*? (A solution of iodine in alcohol.)
4. What should happen if you cover the top of the cylinder and allow all three layers to mix? Try it!

Acids and Bases

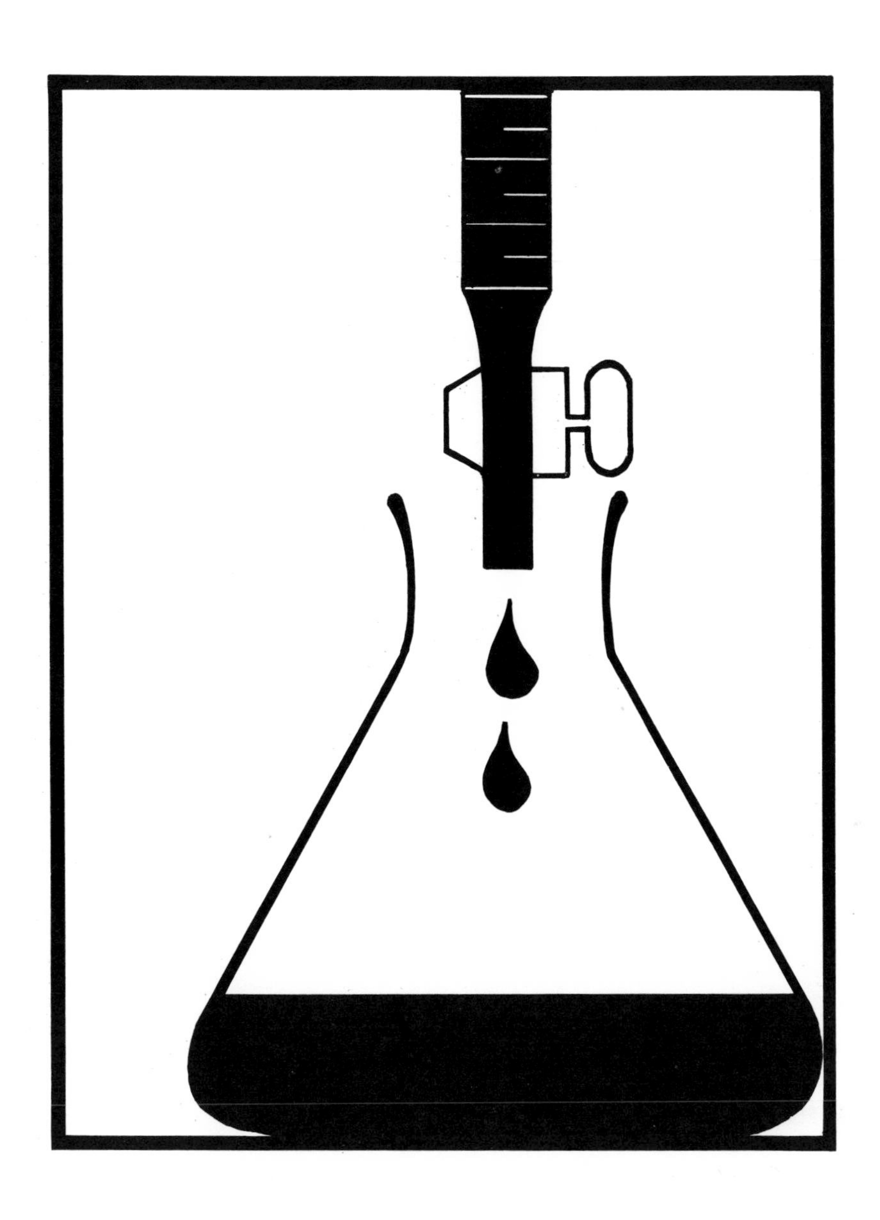

22. Acid–Base Indicators

A few drops of universal indicator are added to a solution. After a few seconds the solution changes color from blue-green to green to yellow to orange to red.

Procedure

1. Prepare a solution by adding 250 mL of water to 250 mL of isopropyl alcohol.
2. To this solution add 1 mL of *tert*-butyl chloride.
3. Add ∿ 4 mL of universal indicator solution.
4. Add 2–3 drops of NaOH solution.
5. Stir vigorously for 5–10 s.
6. Observe color changes.

Reactions

1. Hydrolysis of *tert*-butyl chloride reduces $[OH^-]$ and increases $[H^+]$:

$$CH_3{-}C(CH_3)_2{-}Cl + OH^-(aq) \rightarrow CH_3{-}C(CH_3)_2{-}OH + Cl^-(aq)$$

2. Decreasing the hydroxide concentration causes the indicator to produce the color changes.

Solutions

1. Drug store variety isopropyl alcohol gives poor results because it is a dilute solution. Try various amounts for best results.
2. The NaOH solution is 1.0 M: Dissolve 0.4 g of NaOH in 10 mL of water.

Teaching Tips

NOTES

1. The reaction takes about 2 min.
2. You can repeat the color cycle by adding a little NaOH to neutralize the HCl produced by the reaction.
3. You can also use *tert*-butyl bromide, but use less (about 0.5 mL). It is more rapidly hydrolyzed, so you can expect color changes to appear sooner.

QUESTIONS FOR STUDENTS

1. What were the actual pH changes, initial and final?
2. What was the trend in pH change?
3. Suggest a mechanism for the reaction.
4. Can the reaction be reversed? How?
5. Can the reaction be repeated? How?

23. Acid–Base Indicators: Universal Indicator

A violet-colored solution is poured into a large beaker containing dry ice. The solution is poured back and forth between the two beakers. Violet, blue, green, yellow, and finally orange colors are produced

Procedure

1. Place ~ 100 mL of water in a 500-mL beaker.
2. Add a dropper full of universal indicator solution.
3. Add NaOH solution until a dark, violet color results.
4. Pour this solution into a second large beaker containing about a cup of crushed dry ice.
5. Pour the solution back and forth between the beakers to produce the various colors.

Reactions

1. $$CO_2(g) + H_2O(l) \rightarrow H_2CO_3(aq) \rightleftharpoons H^+(aq) + HCO_3^-(aq)$$

2. The water reacts with the CO_2 to produce an acidic solution. As more H_2CO_3 is formed, the acidity increases and the universal indicator changes to different colors.
3. The following colors are produced at various values: violet, blue, green, yellow, shrimp, and orange at pH 9, 8, 7, 6, 5, and 4, respectively. A few drops of HCl may be necessary to reach pH 4 and produce the orange color.

Teaching Tips

NOTES

1. A little practice is required to get just the right combination of ingredients.
2. You can repeat this demonstration by adding more NaOH solution.

QUESTIONS FOR STUDENTS

1. Why is a range of colors produced in this reaction?
2. Why are sharp, distinct colors produced rather than one color gradually fading into the next?
3. Why was NaOH added at the beginning of the demonstration?

24. Acid–Base Indicators and pH

Eight graduated cylinders are arranged on the demonstration table in four pairs; each cylinder contains a colored solution. A small lump of dry ice is dropped into each of the cylinders. As reactions proceed in each cylinder, various color changes result.

Procedure

1. Arrange eight graduated cylinders, or tall beakers, in pairs.
2. Fill each cylinder about three-fourths with water.
3. Add several drops of the following indicators to each pair of cylinders: pair 1, thymolphthalein; pair 2, phenolphthalein; pair 3, phenol red; and pair 4, bromthymol blue.
4. Add 5–10 mL of NH_4OH to each cylinder.
5. If the cylinder containing thymolphthalein does not turn deep blue, add ammonium hydroxide until it does. Add this same amount of ammonium hydroxide to the other cylinders as well.
6. Add a small piece of dry ice to one member of each pair.
7. Note the changes in color as the CO_2 dissolves and the pH drops. Compare the change in color in each cylinder to the original.

Reaction

As more CO_2 dissolves, the acidity of the solution increases.

Color Changes and pH Range for Indicators

Indicator	Color Range	pH Range
Thymolphthalein	blue to colorless	10.6–9.4
Phenolphthalein	pink to colorless	10.0–8.2
Phenol Red	red to yellow	8.0–6.6
Bromthymol Blue	blue to yellow	7.6–6.0

Solution

The ammonium hydroxide solution is 1 M (see Appendix 2).

Teaching Tips

NOTES

1. The indicators work well within this pH range. You can, of course, add or substitute other indicators.
2. Color changes will occur in sequence if the cylinders are arranged as suggested.

QUESTIONS FOR STUDENTS

1. What pH is indicated by the color change in each cylinder?
2. What does the dry ice do?
3. Write a chemical equation to show the reaction.
4. What causes the bubbling (effervescing)?
5. Does the bubbling action have any effect on the reaction?

25. Acid–Base Indicators: A *Voice-Activated* Chemical Reaction

A flask containing a colored solution is passed around the class. Each student is invited to remove the stopper, speak into the flask, and politely request the color to change to yellow. After 8–10 students have tried, the color of the solution will suddenly change!

Procedure

1. Prepare a flask according to the directions prior to class.
2. Announce to the class that this reaction can be activated by the voice, if a person has just the right voice!
3. Each student is to remove the stopper, speak to the solution, stopper the flask, and give it a quick swirl.

Reactions

1. Eventually, CO_2 from the students' breath will produce enough acid in the solution to cause the color of the indicator to change:

$$CO_2(g) + H_2O(\ell) \rightarrow H_2CO_3(aq) \longleftrightarrow H^+(aq) + HCO_3^-(aq)$$

2. CO_2 also reacts with NaOH. This reaction produces less basic Na_2CO_3:

$$2NaOH(aq) + CO_2(g) \rightarrow Na_2CO_3 + H_2O$$

Solutions

Either of these solutions will work well:

1. Place about 250 mL of 95% ethyl alcohol in a 500-mL Florence flask. Add 5-6 drops of thymolphthalein indicator to the alcohol. Add just enough dilute NaOH to produce a blue color. Stopper the flask until it is used.
2. Prepare the solution as directed in step 1, except use 1–2 drops of phenol red in 250 mL of water. Add 1 drop of a 1 M NaOH solution to produce a red solution. Phenol red is red at pH 8.5 and turns yellow at pH 6.8.

Teaching Tips

NOTES

1. Swirling speeds up the reaction. If you have a large class, omit this step.
2. Be sure to protect the solution from CO_2 in the air prior to the demonstration.

QUESTIONS FOR STUDENTS

1. What chemistry is occurring when you speak into the flask?
2. Why doesn't the color change with the first student?
3. What must you consider when choosing an indicator for this demonstration?
4. Why is a drop of NaOH added prior to the demonstration?

Energy Changes

26. Endothermic Reaction: Ammonium Nitrate

A small amount of ammonium nitrate is added to 100 mL of water. As the solid dissolves, the temperature drops significantly.

Procedure

1. Place ∿ 100 mL of water in a large beaker and record the temperature.
2. Quickly dump 10–15 g of ammonium nitrate into the water.
3. Note the change in temperature as the solid dissolves.

Reaction

The dissolving of ammonium nitrate in water is *endothermic*. The temperature should decrease 6 to 9 °C:

$$\text{heat} + NH_4NO_3(s) + H_2O(\ell) \rightarrow NH_4^+(aq) + NO_3^-(aq)$$

Teaching Tips

NOTES

1. Theoretically, 600–900 calories are absorbed per 100 mL water in this reaction.
2. You will get a more accurate reading by using styrofoam cups. The advantage of using a beaker is that it can be passed around the class for the students to observe.
3. Should you wish to calculate the heat change, 100 mL is a convenient volume.

QUESTIONS FOR STUDENTS

1. Can you think of any practical use for such a reaction (e.g., emergency cold packs)?
2. Why does the temperature drop?
3. Would the temperature drop more if twice as much solid is added? Try it!

27. Endothermic Reaction: Two Solids

Two solids are placed in a beaker and stirred. The beaker is placed in a pool of water on a wooden block. In a few seconds, the beaker gets so cold that it freezes to the block.

Procedure

1. Put ~ 20 g of barium hydroxide crystals [$Ba(OH)_2 \cdot 8H_2O$] in a small (50 mL) beaker.
2. Add 10 g of ammonium thiocyanate to the beaker.
3. Stir the two solids together with a wooden splint.
4. Place the beaker on a small, wooden block with a small pool of water between the beaker and the block.
5. After a couple of minutes, the beaker will freeze to the block.

Reaction

$$\text{heat} + Ba(OH)_2 \cdot 8H_2O(s) + 2NH_4SCN(s) \rightarrow Ba(SCN)_2(s) + 2NH_3(g) + 10H_2O(\ell)$$

Teaching Tips

NOTES

1. You must use barium hydroxide *crystals* [$Ba(OH)_2 \cdot 8H_2O$].
2. Ammonium chloride (~ 7 g) or ammonium nitrate (10 g) may replace ammonium thiocyanate.
3. This demonstration is an excellent way to show that heats of reaction can occur without the presence of a solution.
4. Try varying the amounts of chemicals and the size of the beaker for maximum effect.

QUESTIONS FOR STUDENTS

1. What is water of hydration? Is it important in this reaction?
2. Why would heat be absorbed when water molecules are removed from hydrated barium hydroxide?
3. How does this reaction occur without the compounds being in solution?
4. Does the size of the particles of the two solids matter?

28. Endothermic Reaction: Thionyl Chloride and Cobalt Sulfate

Two chemicals are stirred together in a beaker. The solution is pink, becomes blue, and then becomes very cold.

Procedure

1. DO THIS IN A HOOD.
2. Place about 60 mL of thionyl chloride in a 250-mL beaker.
3. Add 20 g of cobalt sulfate.
4. Stir the two chemicals together. Notice that the pink color changes to blue.
5. Observe the vigorous reaction and the production of a gas.
6. After the reaction has stopped, note the decrease in temperature.

Reaction

1. The gases produced in the reaction are primarily HCl and SO_2:

$$\text{heat} + \underset{\text{(pink)}}{CoSO_4 \cdot 7H_2O(s)} + 7SOCl_2 \rightarrow \underset{\text{(blue)}}{CoSO_4(s)} + 7SO_2(g) + 14HCl(g)$$

2. The temperature drop should be $\sim$ 20 °C.

Teaching Tips

NOTES

1. This endothermic reaction has one nice advantage over others—the low temperature should last for about 10 min.
2. You can also easily freeze this beaker to a wet wooden block.

QUESTIONS FOR STUDENTS

1. What compound is responsible for the pink color? The blue?
2. Why is this reaction performed in a hood?
3. What happened to the water of hydration on the cobalt sulfate molecule?

29. Exothermic Reaction: Calcium Chloride

Solid calcium chloride is added to water, and the temperature of the water increases.

Procedure

1. Quickly dump 10-15 g of calcium chloride into a beaker containing 100 mL of water. (Note the temperature of the water first!)
2. Record the increase in temperature.

Reaction

1. The exothermic heat of solution for calcium chloride is 117 calories per 100 mL of water:

$$CaCl_2(s) + H_2O(\ell) \rightarrow Ca^{2+}(aq) + 2Cl^-(aq) + \text{heat}$$

2. The temperature should increase about 12 °C.

Teaching Tips

NOTES

1. Use a styrofoam cup for a more accurate measurement of the temperature change.
2. Barium oxide can be used instead of calcium chloride.

QUESTIONS FOR STUDENTS

1. Write the chemical equation for this reaction.
2. Why does the temperature increase?
3. Why do some substances absorb heat while others liberate heat when added to water?
4. What is *heat of solution*?

30. Exothermic Reaction: Sodium Sulfite and Bleach

Two solutions are mixed, and a significant increase in temperature results.

Procedure

1. Place 50 mL of laundry bleach in a 250-mL beaker. Record the temperature.
2. Add 50 mL of sodium sulfite solution.
3. Note the increase in temperature.

Reaction

This reaction is highly exothermic; the temperature should increase ~ 20 °C.

$$SO_3^{2-}(aq) + 2OCl^-(aq) \rightarrow SO_4^{2-}(aq) + 2Cl^-(aq) + heat$$

Solutions

1. The sodium sulfite solution is 0.5 M: 6.3 g of Na_2SO_3 per 100 mL of solution.
2. Laundry bleach is a 5.25% solution of sodium hypochlorite, NaOCl.

Teaching Tips

NOTES

1. Use a styrofoam cup for more accurate temperature change measurement.
2. Laundry bleach is a 5.25% solution of sodium hypochlorite, NaOCl.

QUESTIONS FOR STUDENTS

1. Write the chemical equation for this reaction.
2. Would the initial temperature of the solutions affect the final change in temperature? Try it!
3. Would changing the concentration of one of the solutions affect the final change in temperature? Try it!
4. Is this a *redox* reaction? If so, what was reduced and what was oxidized?
5. Could this demonstration be used to determine the *purity* of laundry bleach?

Equilibrium

31. Equilibrium and LeChatelier's Principle

'his demonstration involves the equilibrium reaction, $Fe^{3+}(aq) + SCN^-(aq) \longleftrightarrow FeSCN^{2+}(aq)$. By using petri dish on an overhead projector, you can show the color changes resulting from changes in the oncentration of reactants.

'rocedure

1. Place a petri dish on a clear plastic sheet on an overhead projector.
2. Cover the bottom of the petri dish with the KSCN solution. To indicate the ions present, write $K^+(aq)$ and $SCN^-(aq)$ on the plastic sheet.
3. Add 2–3 drops of $Fe(NO_3)_3$ solution. Note the color change. Write $Fe^{3+}(aq)$ and $NO_3^-(aq)$ to show the ions that were added.
4. Because the color change indicates the formation of a new species, show students that this reaction must proceed as follows:

$$Fe^{3+}(aq) + SCN^-(aq) \rightleftharpoons FeSCN^{2+}(aq)$$

5. Add a small crystal of KSCN to the dish. Do not stir. Notice the formation of a darker color from $FeSCN^{2+}(aq)$. The darker color represents a shift of equilibrium to the right.
6. Add a drop of $Fe(NO_3)_3$ solution. Notice that the color again intensifies, and a shift to the right is indicated.
7. Explain to the students that you can remove some Fe^{3+} by complexing it with Na_2HPO_4. Add a small crystal of Na_2HPO_4; note the immediate clearing of color. A shift of equilibrium to the left is thus indicated.

Reactions

1. The additional $SCN^-(aq)$ from KSCN increases the concentration of reactants and causes a shift of equilibrium toward the products.
2. The additional $Fe^{3+}(aq)$ from $Fe(NO_3)_3$ increases the concentration of reactants and causes a shift of equilibrium toward the right.
3. Adding Na_2HPO_4 reduces the concentration of $Fe^{3+}(aq)$ by forming the colorless complex, $FeHPO_4^+(aq)$. This reaction causes a shift of equilibrium to the left and the formation of a lighter color.

Solutions

1. The KSCN solution is 0.002 M: Dissolve 0.19 g of KSCN per liter.
2. The $Fe(NO_3)_3$ solution is 0.2 M: Dissolve 8 g of $Fe(NO_3)_3 \cdot 9H_2O$ per 100 mL of water. Store this solution in a dropper bottle; very little is used.

Teaching Tips

NOTES

1. You can either do this demonstration in beakers, or project it in a petri dish.
2. Have students make predictions before you perform each part of the demonstration.

QUESTIONS FOR STUDENTS

1. Why are K^+(aq) and NO_3^-(aq) ions not included in the equilibrium reaction?
2. When a crystal of KSCN was added, the solution became darker as a result of the formation of more $FeSCN^{2+}$(aq). How could more $FeSCN^{2+}$(aq) form if no additional Fe^{3+}(aq) was added?
3. What would you suggest be done to restore an equilibrium system after the last part of the demonstration (adding Na_2HPO_4)?

32. Effect of Temperature Change on Equilibrium: Cobalt Complex

An equilibrium system involving the dehydrated–hydrated cobalt complex is produced. When this system is heated, a color change from pink to blue indicates a shift of equilibrium to the right. When the solution is cooled, the color change from blue to pink indicates a shift to the left.

Procedure

1. Place 100 mL of $CoCl_2$ solution in a 250-mL beaker.
2. Add concentrated HCl until the solution changes from pink to blue.
3. Divide the solution into three smaller beakers and treat them as follows:

   a. Place one beaker on a hot plate.
   b. Place one beaker in an ice bath.
   c. Leave one beaker at room temperature as a standard.

4. After a few minutes, show that the heated sample has turned a darker blue and that the cooled sample has turned a light pink.

Reactions

1. This reaction involves the following equilibrium:

$$\text{heat} + [Co(H_2O)_6]^{2+}(aq) + 4Cl^-(aq) \rightleftharpoons [CoCl_4]^{2-}(aq) + 6H_2O$$

(pink) (blue)

2. Addition of heat causes a shift of equilibrium toward products, the blue solution.
3. Cooling causes a shift of equilibrim to the left, the pink hydrated complex.

Solutions

1. The $CoCl_2$ solution is 0.4 M: Dissolve 5.2 g per 100 mL of water.
2. The HCl solution is concentrated.

Teaching Tips

NOTES

1. As indicated in the equation, you may have to add quite a bit of HCl to get the formation of the blue complex.
2. The blue color is due to the tetrachlorocobalt(II) complex, and the pink color is due to the hexaaquacobalt(II) complex.

QUESTIONS FOR STUDENTS

1. Write an equation for the equilibrium system.
2. Why was it necessary to add HCl to establish equilibrium?
3. How does heating shift the equilibrium?
4. What do you think would happen to the equilibrium system if water is added? Try it!

33. Effects of Concentration and Temperature on Equilibrium: Copper Complex

An equilibrium system involving the blue Cu^{2+} and green copper complex is prepared. Changes are made in the concentration of reactants and the temperature to shift the equilibrium.

Procedure

1. Place about 150 mL of $CuSO_4$ solution in a flask. Note the blue color.
2. Add 50 mL of KBr solution. Note the color change from blue to green.
3. Divide the solution into two equal parts for demonstrating the effect of concentration and the effect of temperature.
4. Show the effect of changing concentration: Add a little solid Na_2SO_4, and note the color change from green to blue. Add concentrated HCl to the solution, and note the color change from blue to green.
5. Show the effect of changing temperature: Place beaker from step 4 in an ice bath, and note change of color from green to blue. Remove from ice bath, heat the beaker, and note change of color from blue to green.

Reactions

1. The equilibrium reaction is as follows:

$$\text{heat} + \underset{\text{(blue)}}{CuSO_4(aq) + 4KBr\ (aq)} \rightleftharpoons \underset{\text{(green)}}{K_2[CuBr_4]\ (aq) + K_2SO_4\ (aq)}$$

or, simply:

$$Cu^{2+}\ (aq) + 4Br^-(aq) \rightleftharpoons CuBr_4^{\ 2-}(aq)$$

2. Adding KBr shifts the equilibrium to the right and forms more green complex.
3. Adding Na_2SO_4 forms a basic solution and shifts the equilibrium to the left, forming more blue Cu^{2+}; adding H^+ shifts equilibrium to the right (green).
4. Heating forces the equilibrium to the right; cooling pushes the reaction to the left.

Solutions

1. The KBr solution is saturated: 50 g in 50 mL of hot water. Solid KBr can also be used.
2. The $CuSO_4$ solution is 0.2 M: 50 g of $CuSO_4 \cdot 5\ H_2O$ per liter of water.
3. The HCl is concentrated.

Teaching Tips

NOTES

1. These colors project well and can be demonstrated with petri dishes and an overhead projector.
2. Don't get involved, at this point, with the structure of the copper complex.

QUESTIONS FOR STUDENTS

1. Is this an endothermic or exothermic reaction?
2. How does heating shift the equilibrium?
3. How does adding KBr shift the equilibrium?
4. How does adding Na_2SO_4 cause the equilibrium to shift?

34. Effect of Concentration on Equilibrium: Cobalt Complex

Changing the concentration of reactants results in a shift of equilibrium between pink and blue complex ions of cobalt.

Procedure

1. Shift of equilibrium to the right: Place 20 mL of cobalt(II) chloride solution in a small beaker. Slowly add 40 mL of concentrated HCl. Note the formation of blue color.
2. Shift of equilibrium to the left: Use half of the solution from step 1; save the other half. Add 20 mL of distilled water. Note the color change to pink.
3. Shift of equilibrium to the left: To the remaining solution from step 1, add silver nitrate solution dropwise until a precipitate forms. Note the formation of a pink color.

Reactions

1. This demonstration involves the following equilibrium:

$$[Co(H_2O)_6]^{2+}(aq) + 4Cl^-(aq) \rightleftharpoons [CoCl_4]^{2-}(aq) + 6H_2O(\ell)$$

(pink) (blue)

2. Excess Cl^- causes the formation of more blue tetrachlorocobalt(II) complex.
3. Excess water shifts the equilibrium to the left and forms more pink hexaaquacobalt(II) complex.
4. Silver nitrate removes Cl^-, and the precipitate silver chloride is formed. This precipitation causes the equilibrium to shift to the left, and more pink complex is formed.

Solutions

1. The cobalt(II) chloride solution is 0.2 M: 26 g of $CoCl_2$ per liter of water.
2. The hydrochloric acid is a concentrated solution.
3. The silver nitrate solution is 0.1 M: 1.7 g of $AgNO_3$ per 100 mL of water.

Teaching Tips

NOTES

1. This demonstration projects well. Use petri dishes on an overhead projector.
2. Be sure to use concentrated HCl in step 1. In addition to adding Cl^-, HCl has a dehydrating effect.
3. Have students make predictions of the effect on equilibrium before each part of the demonstration.

QUESTIONS FOR STUDENTS

1. What is the effect of adding a *common ion* to a system in equilibrium?
2. Explain what happened in each part of the demonstration.

3. What might be another way to shift equilibrium to the right?
4. Blue colors are usually associated with hydrated compounds. Why does the hydrated cobalt complex have a pink color?

35. Equilibrium: The Chromate–Dichromate System

Changes in the concentration of reactants and products cause a shift in equilibrium in the yellow chromate–orange dichromate system.

Procedure

EFFECT OF ADDING H^+ AND OH^-

1. Place a petri dish containing chromate solution and another containing dichromate solution on an overhead projector. These solutions will serve as color standards.
2. In a third petri dish, add enough K_2CrO_4 to cover the bottom of the dish.
3. Add HNO_3, dropwise.
4. Note change in color to orange. Ask students what has happened to the equilibrium.
5. Add NaOH solution, dropwise, to the dish. Note formation of the yellow color. Ask students for an explanation.

EFFECT OF CHANGING CHROMATE-DICHROMATE CONCENTRATION

1. Place another petri dish containing K_2CrO_4 on the projector.
2. Add 2–3 drops of NaOH solution.
3. Add barium nitrate solution, dropwise.
4. Note the formation of a precipitate.
5. Add HCl, dropwise, to the solution.
6. Note that the precipitate dissolves and the solution becomes orange. Ask students for an explanation.
7. Replace this petri dish with one containing $K_2Cr_2O_7$ solution: add 2–3 drops of HCl.
8. Add barium nitrate solution, dropwise.
9. Note that no precipitate forms.
10. Add NaOH solution, dropwise, to the solution.
11. Notice that a precipitate forms and the color changes to yellow.

Reactions

1. This demonstration illustrates the following equilibrium:

$$\underset{\text{(yellow)}}{2CrO_4^{2-}(aq) + 2H^+(aq)} \longleftrightarrow \underset{\text{(orange)}}{Cr_2O_7^{2-}(aq) + H_2O(\ell)}$$

2. Adding acid increases H^+ and shifts the equilibrium to the right.
3. Adding base decreases H^+ and shifts the equilibrium to the left.
4. Adding barium nitrate to chromate precipitates barium chromate, an insoluble compound:

$$Ba^{2+}(aq) + CrO_4^{2-}(aq) \rightarrow BaCrO_4(s)$$

5. Adding H^+ favors the formation of $Cr_2O_7^{2-}$; therefore, adding HCl dissolves the precipitate and shifts the equilibrium to the right.

6. Adding barium nitrate to dichromate does not form a precipitate because barium dichromate is soluble.
7. $BaCr_2O_7$ is more soluble than $BaCrO_4$. Thus, adding OH^- favors the formation of more CrO_4^{2-}, and the precipitate forms.
8. Summary of Reaction Equilibrium:

$$2OH^- + Cr_2O_7^{2-} \rightarrow 2CrO_4^{2-} + H_2O$$

$$2H^+ + 2CrO_4^{2-} \rightarrow Cr_2O_7^{2-} + H_2O$$

$$2H^+ + 2OH^- \rightarrow 2H_2O$$

Solutions

1. The potassium chromate solution is 0.1 M: 19 g of K_2CrO_4 per liter.
2. The potassium dichromate solution is 0.1 M: 29 g of $K_2Cr_2O_7$ per liter.
3. The barium nitrate solution is 0.1 M: 26 g of $Ba(NO_3)_2$ per liter.
4. The HNO_3 and HCl solutions are 1 M (see Appendix 2).
5. The sodium hydroxide solution is 1.0 M: 40 g of NaOH per liter.

Teaching Tips

NOTES

1. This demonstration involves a great deal of chemistry! Be sure that students are following each step.
2. Prepare small amounts of solutions and store them in dropper bottles.

QUESTIONS FOR STUDENTS

1. Write the chemical equation for the equilibrium reaction.
2. Write an equation to show each shift in equilibrium.
3. Will any acid cause the equilibrium to shift? Any base? (Try it!)
4. What have you learned about the solubility of barium chromate and barium dichromate?
5. Explain the overall reaction.

36. Equilibrium in the Gas Phase

A reddish-brown gas is prepared and placed in two small test tubes. When one tube is placed in boiling water, the color of the gas changes to a deep brown color. When the other tube is placed in an ice bath, the gas becomes almost colorless. When both tubes are allowed to reach room temperature, the gas in both again becomes reddish-brown.

Procedure

PREPARING THE GAS TUBES

1. Prepare a gas generator by attaching a rubber tube to a small side-arm flask.
2. IN A HOOD, add 10 mL of concentrated nitric acid to the flask.
3. Drop in a copper penny.
4. A deep-red gas, NO_2, will immediately form. Allow enough of the gas to form to displace the air in the flask and two test tubes. Colorless NO forms first. Colorless bubbles rise to the surface where they mix with O_2 in the air and immediately form NO_2.
5. Fill the two tubes with the brown gas. Stopper the tubes.
6. Stop the reaction in the flask by filling the flask with water. Notice the blue color of the solution and the small size of the penny.

DEMONSTRATING EQUILIBRIUM

1. Point out the color of the gas in the tube. Put the equilibrium reaction on the board.
2. Place one tube in a beaker of boiling water. Notice the change in color.
3. Place the other tube in an ice bath. Notice the formation of a colorless gas.
4. Remove both tubes and allow them to come to room temperature. Note the restoration of the brown color in both tubes.

Reactions

1. The equations for the production of the gas are as follows:

$$3Cu + 8H^+ + 2NO_3^- \rightarrow 3Cu^{2+} + 2NO + 4H_2O$$

$$2NO + O_2 \rightarrow 2NO_2$$

2. The equilibrium mixture in the tubes consists of NO_2 and N_2O_4. They react according to the equation:

$$\underset{\text{(red)}}{2NO_2(g)} \rightleftharpoons \underset{\text{(colorless)}}{N_2O_4}$$

3. When the equilibrium mixture is heated, the equilibrium shifts toward the formation of brown NO_2.
4. When the mixture is cooled, the equilibrium shifts toward the formation of more colorless N_2O_4.

Teaching Tips

NOTES

1. The preparation of the gas tubes must be done IN A HOOD: NO_2 is TOXIC. It has been estimated that five pennies could produce enough poisonous NO_2 to fill a laboratory!
2. Fat, short, clear plastic test tubes work best. They can be used for several weeks, if tightly stoppered.

QUESTIONS FOR STUDENTS

1. Write the chemical equation for the production of the gas.
2. Why is the solution formed in the reaction vessel blue?
3. Could the chemical change in the tubes be due to an increase in pressure, rather than an increase in temperature?
4. Devise an experiment to prove or disprove your hypothesis.
5. Is this an exothermic or endothermic reaction?

37. Equilibrium: Temperature and the Ammonium Hydroxide–Ammonia System

A large test tube containing a pink solution is heated and the color disappears. The test tube is cooled in tap water and the pink color returns.

Procedure

1. Place about 400 mL of water in a large beaker.
2. Add a few drops of phenolphthalein indicator.
3. Add 1 drop of concentrated ammonia. The solution should be pink, but not dark pink.
4. Fill a large test tube one-half full with this solution.
5. Heat the tube over a burner. Note that the pink color fades and disappears.
6. Let the tube cool for a few seconds, then place it in a stream of cold water. Notice that the pink color returns.

Reactions

1. The reaction is basically an equilibrium between the ammonium hydroxide and the nonionized ammonia:

$$NH_4^+(aq) + OH^-(aq) \rightleftharpoons NH_3(aq) + H_2O(\ell)$$

2. Heat shifts the equilibrium and reduces the ammonium hydroxide so that it no longer reacts with the indicator.

Teaching Tips

NOTES

1. If you add too much ammonia, the pink color will persist when the tube is heated.
2. You can do this demonstration in a larger container, but it will take longer to cool and restore the color.

QUESTIONS FOR STUDENTS

1. Write the reaction for this equilibrium.
2. Is the reaction endothermic or exothermic?
3. Why does the demonstration not work well if too much ammonia is added?

38. Equilibrium: Effect of Temperature

When a test tube containing a red solution is heated, the solution turns blue. The tube is then partly immersed in an ice–salt bath. The solution in the immersed part of the tube turns pink, but the top part remains blue.

Procedure

1. Heat 150 mL of cobalt(II) chloride solution in a flask until the red solution turns blue.
2. Fill a large test tube with the blue solution and immerse it half-way into a large beaker containing crushed ice and salt.
3. Notice that the solution in the bottom part of the tube turns pink.

Reactions

1. The equilibrium reaction is as follows:

$$\underset{\text{(blue)}}{CoCl_4^{2-}(aq)} + 6H_2O(\ell) \rightleftharpoons \underset{\text{(pink)}}{[Co(H_2O)_6]^{2+}(aq)} + 4Cl^-(aq) + \text{heat}$$

2. Heating forces the equilibrium to the left, and more of the blue complex is formed. Cooling causes formation of the pink complex ion.
3. Cobalt(II) chloride, $CoCl_2 \cdot 6H_2O$, is red in solution. The hydrated complex ion, $[Co(H_2O)_6]^{2+}$, is pink.

Solution

The cobalt chloride solution is saturated.

Teaching Tips

NOTES

1. This demonstration is unusual because it allows you to produce pink hydrated and blue dehydrated forms in the same container.
2. This reaction is the basis for the chemical weather predictor.

QUESTIONS FOR STUDENTS

1. What is the temperature of the ice–salt bath? (~ -18 °C)
2. What effect does cooling have on the equilibrium system?
3. Write an equation for this reaction.
4. What is the chemical name for the hydrated ion? [hexaaquacobalt(II)]

39. Effect of Pressure on Equilibrium

A side-arm flask containing a colorless solution is connected to an aspirator on a water faucet. When the water faucet is turned on, the solution begins to bubble and later turns pink.

Procedure

1. Fit a 250-mL side-arm flask with a tight-fitting rubber stopper and vacuum tubing connected to the aspirator on a laboratory water faucet.
2. Add 150 mL of $NaHCO_3$ solution to the flask.
3. Add 3–4 drops of phenolphthalein indicator. If the solution is pink, add dilute HCl dropwise until the color just disappears.
4. Stopper the flask and turn on the water.
5. Notice the formation of a gas and the eventual color change to pink.

Reactions

1. The equilibrium system is represented by the following scheme:

$$HCO_3^-(aq) \rightleftharpoons OH^-(aq) + CO_2(g)$$

2. Decreasing pressure forces the equilibrium to the right, and causes the release of CO_2 and a more alkaline solution.
3. The alkaline solution causes the indicator to become pink.

Solution

The $NaHCO_3$ solution is saturated: Dissolve 32 g in 200 mL of water.

Teaching Tips

NOTES

1. Students should know that phenolphthalein is colorless in acid solution and pink in alkaline solution.
2. In addition to reducing the pressure, the vacuum is removing a product (CO_2).

QUESTIONS FOR STUDENTS

1. Write the equilibrium reaction for this system.
2. Why does the solution become pink?
3. Which is more significant in this change, reducing pressure or removing CO_2?

40. Effect of Hydrolysis on Equilibrium

A beaker contains a clear solution. When water is added to the solution, a dense, white precipitate forms. When HCl is added, the precipitate dissolves and the solution becomes clear again.

Procedure

1. Place about 250 mL of water in a beaker.
2. Add a small amount (one or two small lumps) of antimony trichloride ($SbCl_3$). Note the cloudy solution.
3. Add concentrated HCl until the solution just clears.
4. Add water until a precipitate again appears and the solution becomes cloudy.

Reactions

1. The equilibrium system is as follows:

$$\underset{\text{(clear)}}{SbCl_3(s) + H_2O(\ell)} \rightleftharpoons \underset{\text{(cloudy)}}{SbOCl\,(s) + 2HCl(aq)}$$

2. Adding HCl shifts the equilibrium to the left and more clear antimony trichloride solution is produced.
3. Adding water shifts the equilibrium to the right and more white antimony(III) oxychloride, SbOCl, is formed.

Teaching Tips

NOTES

1. This reaction projects well. Use petri dishes and an overhead projector.
2. Use the smallest amount of $SbCl_3$ possible to give a cloudy solution. Otherwise, a large volume of HCl will be needed to clear the solution.
3. You can also do this demonstration with bismuth chloride; bismuthyl chloride is formed as a hydrolysis product:

$$\underset{\text{(clear)}}{BiCl_3(s) + H_2O(\ell)} \rightleftharpoons \underset{\text{(white)}}{BiOCl(s) + 2HCl(aq)}$$

QUESTIONS FOR STUDENTS

1. What is hydrolysis?
2. How does antimony trichloride change when water is added?
3. Would addition of a chloride rather than HCl (ammonium chloride, for example) cause the equilibrium to shift. Try it!
4. Can you think of other things to do to cause a shift in equilibrium? Try them!

41. Solubility Product: Effect of Concentration

A saturated solution of lead bromide is prepared. When a solution of lead nitrate is added, no precipitate forms. However, when a solution of sodium bromide is added, extensive precipitation occurs.

Procedure

1. Place about 150 mL of lead bromide (see solutions below) in two beakers.
2. In one beaker add 10–15 mL of lead nitrate. Notice that no precipitate forms.
3. In the other beaker, add the same amount of sodium bromide solution. Notice that precipitation is heavy.

Reaction

$$PbBr_2(s) \rightleftharpoons Pb^{2+}(aq) + 2Br^-(aq)$$

This reaction clearly shows that the concentration of Br^- enters the solubility product expression for lead bromide to a higher power than does the concentration of lead.

$$K_{sp} = [Pb^{2+}]\,[Br^-]^2 = 6.3 \times 10^{-6}$$

Solutions

1. The lead nitrate solution is 1.0 M: 331 g of $Pb(NO_3)_2$ per liter.
2. The sodium bromide solution is 1.0 M: 103 g of NaBr per liter.
3. Prepare the lead bromide as follows: Add 100 mL of 1.0 M lead nitrate solution to 200 mL of 1.0 M sodium bromide solution. Filter to remove the precipitated lead bromide. You may need to use suction filtration.
4. Prepare a saturated solution of lead bromide: 10 g of $PbBr_2$ in 200 mL of water.

Teaching Tips

NOTES

1. Use the lead bromide solution sparingly.
2. Lead bromide is only slightly soluble. Use warm water.
3. This demonstration projects well. Use petri dishes and an overhead projector.

QUESTIONS FOR STUDENTS

1. What is a solubility product constant?
2. Write the expression for lead bromide.
3. Why does additional Pb^{2+} cause no precipitation of lead bromide, whereas additional Br^- does?
4. How is equilibrium involved in this reaction?

Kinetics

42. Carbon as a Catalyst

Try to burn a cube of sugar held on a toothpick. It doesn't burn, it only melts when heated with a match. After you dab a second sugar cube into a pile of cigarette ashes, it can be burned when lit with a match.

Procedure

1. Show your students that you cannot burn a cube of table sugar held on a toothpick (or with tongs) with the heat from a match.
2. Take another cube, dab it in a pile of cigarette ashes (cover at least two sides).
3. Now, when the cube is heated with a match, the cube will burn.

Reaction

The carbon in the cigarette ashes acts as a catalyst in the combustion of sugar.

Teaching Tips

NOTES

1. Fine carbon will work as well as cigarette ashes.
2. The melting point of sucrose is 185 °C.

QUESTIONS FOR STUDENTS

1. What is a catalyst?
2. How does carbon act as a catalyst?
3. What other substances can you think of that act as catalysts?

43. Medicine Cabinet Kinetics: How Fast Is the Fizz?

Fresh *Alka Seltzer* tablets are dropped, simultaneously, into beakers of cold water and warm water. The one dropped in warm water fizzes and reacts much faster than the one dropped in cold water.

Procedure

1. Place ~ 150 mL of warm water in a beaker.
2. Place the same amount of cold water in a second beaker.
3. Drop an *Alka Seltzer* tablet into each beaker.
4. Note the rate of reaction in each beaker.

Reactions

1. *Alka Seltzer* consists of a mixture of the following compounds: calcium dihydrogen phosphate, $Ca(H_2PO_4)_2$; citric acid; aspirin; and sodium bicarbonate, $NaHCO_3$.

2. Calcium dihydrogen phosphate is a source of hydrogen ion:

$$Ca(H_2PO_4)_2(s) \rightarrow 2H^+(aq) + 2(HPO_4)^{2-}(aq) + Ca^{2+}(aq)$$

3. Sodium bicarbonate is a source of bicarbonate ion, HCO_3^-:

$$NaHCO_3(s) \rightarrow Na^+(aq) + HCO_3^-(aq)$$

4. H^+, HCO_3^-, and H_2O react to produce carbon dioxide and water:

$$H^+(aq) + HCO_3^-(aq) + H_2O(\ell) \rightarrow CO_2(g) + 2H_2O(\ell)$$

5. The overall reaction is

$$Ca(H_2PO_4)_2\ (s) + 2NaHCO_3(s) \rightarrow 2CO_2(g) + 2H_2O(\ell) + CaHPO_4(s) + Na_2HPO_4(s)$$

Teaching Tips

NOTES

1. Be sure to use FRESH *Alka Seltzer* tablets.
2. This reaction is the same that occurs when baking powder, which also contains calcium dihydrogen phosphate and sodium bicarbonate, decomposes and causes baked products to rise.

QUESTIONS FOR STUDENTS

1. What generalization about rates of reactions can be drawn from this demonstration?
2. If the *Alka Seltzer* tablets are old, they don't produce much fizz. Why?
3. How would the rate of reaction at room temperature compare with that in cold water? Hot water? Try these!

44. Catalytic Decomposition of Hydrogen Peroxide: Foam Production

A tremendous amount of foam shoots from a graduated cylinder when detergent and potassium iodide are added to hydrogen peroxide.

Procedure

1. Place a large graduated cylinder (500 mL) in a plastic tray or in a laboratory sink.
2. Pour ∼ 50 mL of 30% hydrogen peroxide into the cylinder. (CAUTION!)
3. Add a squirt of dishwashing detergent and a drop of food coloring.
4. Add about one-fourth of a spoonful of solid KI.

Reactions

1. The rapidly catalyzed decomposition of hydrogen peroxide produces oxygen gas, which forms a foam with the liquid detergent:

$$2H_2O_2(aq) \rightarrow 2H_2O(\ell) + O_2(g)$$

2. The actual decomposition of H_2O_2 in the presence of iodide ion occurs in two steps. The first reaction is the rate-determining reaction.

$$H_2O_2(aq) + I^-(aq) \rightarrow H_2O(\ell) + OI^-(aq)$$

$$H_2O_2(aq) + OI^-(aq) \rightarrow H_2O(\ell) + O_2(g) + I^-(aq)$$

Solution

You must use 30% hydrogen peroxide for best results. Do not use the 3% hydrogen peroxide available from a drugstore.

Teaching Tips

NOTES

1. Be careful when using 30% hydrogen peroxide. Wear gloves and avoid contact with this solution.
2. You can also show the decomposition of hydrogen peroxide on an overhead projector. Place a small amount of 3% hydrogen peroxide in a petri dish on an overhead projector. Add a pinch of potassium iodide, or manganese dioxide, and note the evolution of oxygen gas bubbles.
3. The catalytic decomposition of hydrogen peroxide occurs when the 3% solution is placed on a wound. Catalase, an enzyme in the blood, catalyzes the reaction.

QUESTIONS FOR STUDENTS

1. How does a catalyst work?
2. What happened to the KI?
3. How can you account for the large amount of foam produced?
4. What evidence is there that iodine is produced? (The brown color of the foam.)

45. A Catalyst in Action

A catalyst is added to a solution. A green-colored activated complex is formed and the reaction proceeds. When the green complex disappears, the reaction ceases.

Procedure

1. Place 200 mL of sodium potassium tartrate solution in a large (600 mL) beaker.
2. Warm the solution gently to about 70 °C on a hot plate. (CAUTION!)
3. Add 80 mL of hydrogen peroxide.
4. Add a few crystals (about a spoonful) of cobalt(II) chloride.
5. A vigorous reaction will occur. Note the appearance of the green activated complex and the extensive bubbling.

Reactions

1. This reaction involves the oxidation of tartaric acid, ($HO_2CCH(OH)CH(OH)CO_2H$), by hydrogen peroxide in the presence of a cobalt(II) chloride catalyst.
2. The green color is due to the formation of a cobalt–tartrate activated complex.
3. Note that the original catalyst, cobalt(II) chloride, is pink. As the tartrate is oxidized, the activated complex is broken down to the original catalyst, and the pink color returns.
4. Oxygen and carbon dioxide gases are produced. Oxalic acid, HO_2CCO_2H, is probably produced also.

Solutions

1. Sodium potassium tartrate: 12 g of $KNaC_4H_4O_6 \cdot 4H_2O$ per 200 mL of distilled water.
2. Hydrogen peroxide: You must use 6% H_2O_2 for this reaction, NOT the common 3% drugstore variety. Two sources of 6% H_2O_2 are to purchase it at a drugstore (*Clairoxide* is one brand), or to dilute 30% hydrogen peroxide by adding 200 mL of 30% H_2O_2 per liter of solution. CAREFUL with 30% H_2O_2!

Teaching Tips

NOTES

1. This demonstration is one of the few that allows one to actually observe the formation of an activated complex and its action.
2. Do not exceed 70 °C, or the solution may froth and overflow the beaker.
3. This demonstration is also an excellent method to show the relationship between temperature and reaction rate. Typically, an initial temperature of 50, 60, or 70 °C will produce a reaction time of 200, 90, or 40 s, respectively.
4. As a general rule, increasing the temperature of reaction by 10 degrees will double the rate of reaction.

QUESTIONS FOR STUDENTS

1. Write chemical equations for the reactions.
2. What is the formula of the activated complex?
3. What is the role of the hydrogen peroxide?
4. What gases are produced?
5. How can you be sure that the green color is actually due to the activated complex?

46. Autocatalysis

A graduated cylinder contains a blue solution. A few drops of acid are placed on top of the solution, and in a few seconds a yellow layer appears. Within a few minutes, the yellow layer gradually moves down the column as the catalyzed reaction proceeds.

Procedure

1. Make the dilute sulfuric acid solution indicated below.
2. Place 50 mL of water in a beaker.
3. Add 4 g of potassium chlorate, 12.5 g of sodium sulfite, and a small amount (~ 5 mg) of bromophenol blue indicator.
4. In a second beaker, add 4 mL of the sulfuric acid solution to 50 mL of water.
5. Slowly, with constant stirring, add the diluted acid from the second beaker to the solution in the first beaker. Stir until everything dissolves. The solution should be blue-violet.
6. Fill a 100-mL graduated cylinder with the solution.
7. Carefully add two droppers full of the sulfuric acid solution to the top of the liquid in the cylinder.
8. Soon a yellow color will appear at the top of the solution, and a yellow-blue interface will form.
9. Observe for several minutes as the yellow-blue interface moves down the graduated cylinder.

Reactions

1. This reaction is a redox reaction:

$$ClO_3^-(aq) + 3HSO_3^-(aq) \rightarrow Cl^-(aq) + 3SO_4^{2-}(aq) + 3H^+(aq)$$

(blue) (yellow)

2. The pH on the reactant side is about 7; the pH on the product side is less than 7. Thus, the reaction proceeds only in an acidic solution.

3. Dropping sulfuric acid on the surface produces acidic products: $H^+ + SO_2^{2-} \rightarrow HSO_3^-$ These acid products catalyze further reactants to produce additional acidic products; hence, the autocatalytic effect results.

4. Bromophenol blue indicator is yellow in the presence of an acid. Thus, as autocatalysis proceeds, the blue color of the indicator changes to yellow.

5. The blue solution has a pH between 6.5 and 7.0, because of the buffering effect of the bisulfite–sulfite ions.

Solution

Sulfuric acid: Prepare a dilute solution by adding 10 mL of concentrated sulfuric acid to 35 mL of distilled water. *Use this acid only in the above Procedure steps 3 and 6.*

Teaching Tips

NOTES

1. The amount of bromophenol blue indicator is not critical. An amount about the size of a matchhead is about right.

2. For best results, use anhydrous analytical grade chemicals

3. Point out that the dropper full of acid merely begins the reaction. If the color change were due to this small amount of acid alone, the entire contents of the cylinder would immediately turn yellow!

QUESTIONS FOR STUDENTS

1. What is meant by *autocatalysis*?
2. Can you think of other autocatalytic reactions? (Many biochemical reactions are of this type. Pepsinogen is activated by hydrogen ion in the stomach to form pepsin. Pepsin then catalyzes the conversion of pepsinogen to additional pepsin.)
3. In this redox reaction, what is oxidized? What is reduced?
4. What did you observe happening at the interface between the yellow and blue layers?

47. The Starch–Iodine Clock Reaction

Each of two beakers contains a colorless solution. The solutions are mixed by pouring from one beaker to the other. After a few seconds, the mixed solution suddenly turns dark blue. Changing the concentration or the temperature of the solutions changes the time required for the blue color to be produced.

Procedure

1. Place 50 mL of solution *A* in a 250-mL beaker.
2. Place the same volume of solution *B* in a second beaker.
3. Mix the two solutions by pouring from one beaker into the other several times, then hold the filled beaker in full view of the class.
4. Note the time required for a reaction to occur.
5. Repeat, but use solution *A* that has been diluted to one-half concentration. Note the time required for a reaction to occur.
6. Repeat, using solution *A* that has been warmed to 35 °C. Note the time required for a reaction to occur.

Reactions

The mechanism is not completely understood; however, the following simplified sequence is proposed:

1. IO_3^- reacts with HSO_3^- to form I^-:

$$IO_3^- + 3HSO_3^- \rightarrow I^- + 3H^+ + 3SO_4^{2-}$$

2. I^- reacts with IO_3^- to form I_2.
3. I_2 is immediately consumed by reaction with HSO_3^-:

$$I_2 + HSO_3^- + H_2O \rightarrow 2I^- + SO_4^{2-} + 3H^+$$

4. When all of the HSO_3^- has been used up, I_2 accumulates.
5. Iodine reacts with starch to form a colored complex:

$$I_2 + \text{starch} \rightarrow \text{blue-colored complex}$$

Solutions

1. Solution *A*: 4.3 g of KIO_3 per liter.
2. Solution *B*: A starch solution made as follows: Make a paste of 4 g of soluble starch in a small amount of warm water. Slowly add 800 mL of boiling water. Boil for a few minutes then cool the solution. Add 0.2 g of $Na_2S_2O_5$ (sodium metabisulfite). Add 5 mL of 1.0 M sulfuric acid (see Appendix 2). Dilute to 1 liter.

Teaching Tips

NOTES

1. This demonstration allows you to beautifully illustrate the dependence of reaction rates on concentration and temperature.
2. Do not heat the solutions above 40 °C. The starch–iodine complex is unstable above 50 °C. Best results are obtained when the solutions are allowed to stabilize at room temperature for a couple of hours prior to mixing.
3. Ideally, 10–15 s should be required when the solutions are mixed at room temperature. If the reaction is too slow, add a little more sodium metabisulfite or more acid to solution *B*. If the reaction is too fast, dilute solution *A*.
4. $Na_2S_2O_5$ hydrolyzes in solution to $NaHSO_3$.
5. After sulfuric acid is added to solution *B*, it must be used within 10–12 h. If you need to keep the solution longer, add the acid just before using the solution.

QUESTIONS FOR STUDENTS

1. Propose a simple mechanism for this reaction.
2. What is the role of the starch?
3. Why is there a delay before the reaction occurs?
4. What are the effects of concentration and temperature on reaction rates?
5. Why is this called a *clock* reaction?

18. The *Old Nassau* Clock Reaction

Three colorless solutions are mixed. In a few seconds, the solution turns bright orange, then suddenly turns dark blue.

Procedure

1. Label three 250-mL beakers *A*, *B* and *C*.
2. Place 50 mL of solutions *A*, *B*, and *C* into their respective beakers.
3. Mix the solutions IN THIS ORDER: Add *A* to *B* to *C*.
4. Hold the beaker in view of the class.

Reactions

$$IO_3^- + 3HSO_3^- \rightarrow I^- + 3SO_4^{2-} + 3H^+$$

$$Hg^{2+} + 2I^- \rightarrow HgI_2 \text{ (orange)}$$

$$6H^+ + IO_3^- + 5I^- \rightarrow 3I_2 + 3H_2O$$

$$I_2 + \text{starch} \rightarrow \text{(blue)}$$

Solutions

1. Solution *A*: Dissolve 4 g of soluble starch in 500 mL of boiling water. (Make a paste with a few milliliters of water first.) Cool, add 15 g of $NaHSO_3$ and dilute to 1 liter with distilled water.
2. Solution *B*: 3 g of $HgCl_2$ per liter distilled water.
3. Solution *C*: 15 g of KIO_3 per liter distilled water.

Teaching Tips

NOTES

1. To speed up the reaction, use less of solution *B*.
2. The reaction is called the *Old Nassau* reaction because it produces the colors of Princeton University (orange and black). Nassau Hall is one of the older buildings on the Princeton campus.

QUESTIONS FOR STUDENTS

1. Propose a mechanism to explain how this reaction can produce two distinct colors.
2. How can the reaction rate be increased?
3. Is it necessary to mix the solutions in a particular order? Try it!
4. What compound is formed when the solution turns orange?

49. Disappearing Orange Reaction: Now You See It, Now You Don't!

Two beakers, *A* and *B*, contain colorless solutions. A small amount of solution from beaker *A* is poured into beaker *B* and a bright orange color is produced. When the remainder of solution in beaker *A* is poured into *B*, however, the orange color disappears.

Procedure

1. Add equal volumes of solutions *A* and *B* to two beakers labeled *A* and *B*.
2. Pour enough of solution *A* into solution *B* to produce a bright orange color.
3. Pour the remaining solution *A* into the *B* beaker and note that the orange color disappears!

Reactions

$$Hg^{2+} + 2I^- \rightarrow HgI_2 \text{ (orange)}$$

$$HgI_2 + 2I^- \rightarrow HgI_4^{2-} \text{ (colorless)}$$

Solutions

1. Solution *A*: 5 g of potassium iodide in 300 mL of water.
2. Solution *B*: 2 g of mercuric chloride in 300 mL of water.

Teaching Tips

NOTES

1. KI forms an orange precipitate with $HgCl_2$. Excess KI dissolves this precipitate and forms a colorless complex.
2. For a variation of this reaction, see *Old Nassau* reaction.

QUESTIONS FOR STUDENTS

1. Write the chemical equations for the reactions.
2. Is this a redox reaction?
3. What could you do to change the rate of the reaction?
4. How can you explain the disappearance of the orange color?

50. A Traffic Light Reaction

A flask containing a pale yellow solution is gently swirled. The solution turns red. The flask is shaken, and the solution turns green.

Procedure

1. Place 50 mL of solution *A* in a 250-mL flask.
2. Add 5–10 mL of indicator solution.
3. Stopper the flask.
4. At the beginning of the demonstration, the solution should be light yellow.
5. Gently swirl the flask to produce the red color.
6. Give the flask a quick shake to produce the green color.

Reactions

1. The indicator is reduced by alkaline dextrose, and a yellow color is produced.
2. When the flask is swirled, oxygen is added, the indicator is oxidized, and the red color is produced.
3. Shaking the flask introduces even more oxygen and causes further oxidation of the indicator to the green color.
4. Upon standing, the dextrose reduces the indicator back to the yellow color.

Solutions

1. Solution *A*: 3 g of dextrose (glucose) and 5 g of NaOH in 250 mL of water.
2. The indigo carmine indicator is a 1.0% solution. (With practice, you can use a small amount of solid.)

Teaching Tips

NOTES

1. If the red color does not persist, adjust the number of drops of indicator.
2. This traffic light has an advantage over others—a magnetic stirrer is not required.
3. For a variation of this reaction, see Demonstration 59, The Blue Bottle Reaction.

QUESTIONS FOR STUDENTS

1. Propose a chemical equation for the reaction.
2. Is this a redox reaction? If so, what is oxidized and what is reduced?
3. What role does the indicator play?
4. What happens when the flask is swirled? Shaken?
5. Will this reaction *run down* if the stopper remains in the flask?

51. The Quick Gold Reaction

Two colorless solutions are mixed by pouring from one beaker into another. After 15–20 s, the solutio suddenly turns bright yellow-gold.

Procedure

1. Place 100 mL of solution *A* in a beaker. (CAUTION!)
2. Place 100 mL of solution *B* in a second beaker.
3. Pour the contents of beaker *A* into the second beaker.
4. Notice the sudden formation of a yellow-gold color after about 15–20 s.

Reactions

1. Acetic acid and sodium thiosulfate react to produce hydrogen sulfide. This reaction is slow.
2. The hydrogen sulfide reacts with sodium arsenite in a faster reaction to produce yellow arsenious sulfide.

Solutions

1. Solution *A*: Dissolve 2 g of sodium arsenite in 100 mL of water. (CAUTION!) Then add 11 mL of glacial acetic acid.
2. Solution *B*: 20 g of sodium thiosulfate in 100 mL of water.

Teaching Tips

NOTES

1. Sodium thiosulfate is photographer's *hypo*.
2. Glacial acetic acid is 100% acetic acid. HANDLE CAUTIOUSLY.
3. This reaction can be timed to the second; the demonstrator can predict when it will change color.
4. This demonstration is an excellent way to introduce chemical kinetics! Students should predict the formation of an intermediate product.
5. Use the arsenite solution with caution.

QUESTIONS FOR STUDENTS

1. Write the chemical equation for the reactions.
2. What is the gas produced in the reaction?
3. Which is the slow reaction? Which is fast? Which reaction is the rate-determining reaction?
4. Is this a redox reaction? If so, what is the oxidizer and the reducer?

52. An Oscillating Reaction: Clear–Brown

A solution is stirred with a magnetic stirrer. The solution bubbles and fizzes, then turns brown, then colorless, and then brown again. The oscillations will continue for several minutes.

Procedure

1. Place a large beaker on a magnetic stirrer.
2. Add 450 mL of water and 50 mL of concentrated sulfuric acid. (CAREFUL!)
3. While the solution is stirring, add the following: 3 spoons of malonic acid, 2 spoons of $KBrO_3$, and ∿ one-fourth spoon of $MnSO_4$.
4. The solution will fizz, then turn brown.
5. Oscillations will begin after a few seconds.

Reactions

1. Oscillating reactions are very complex reactions. This one is thought to involve over 20 chemical species and 18 steps in the reaction mechanism.
2. Products of this reaction include CO_2; formic acid, HCOOH; and bromomalonic acid, $BrCH(COOH)_2$.
3. Formation of different oxidation states of manganese results in the colorless and brown appearance of the solution as it oscillates.

Teaching Tips

NOTES

1. The amounts of chemicals in this demonstration are not critical, thus we suggest *spoons*.
2. An excellent article on oscillating reactions appears in *Scientific American*, March, 1983, page 112.
3. If a magnetic stirrer is not available, stir the solutions in the beaker with a stirring rod.

QUESTIONS FOR STUDENTS

1. Why is stirring important in this reaction?
2. What causes the effervescence?
3. Why does the reaction eventually stop?
4. What can you suggest that could be done to renew the oscillations? Try it!

53. An Oscillating Reaction: Yellow–Blue

Solutions are mixed and placed on a magnetic stirrer. The color of the solution changes from light yellow to blue to light yellow. This oscillation will continue for 10–15 min.

Procedure

1. Place 100 mL of solution *A* in a 500-mL beaker on a magnetic stirrer.
2. Set the stirrer on its slowest setting.
3. Add 100 mL of solution *B*.
4. Add 100 mL of solution *C*.
5. Oscillations will begin after a few seconds.

Reactions

1. This demonstration involves a series of complex reactions. In the first series of reactions, oxygen gas and iodine are formed.
2. The iodine reacts with the starch to produce the blue color.
3. As the iodine is used up in another series of reactions, the color fades but it is formed again when the concentration of iodine increases.

Solutions

1. Solution *A*: Add 40 mL of 30% hydrogen peroxide to 100 ml of water. (CAUTION!)
2. Solution *B*: While stirring, add 4.3 g of KIO_3 and 0.5 mL of concentrated sulfuric acid to 100 mL of water.
3. Solution *C*: Prepare a paste of 0.15 g of soluble starch in hot water. While stirring, add this to 500 mL of hot water. Then add 7.8 g of malonic acid and 1.7 g of $MnSO_4 \cdot H_2O$.

Teaching Tips

NOTES

1. Notice that 30% hydrogen peroxide must be used. CAREFUL with this solution!
2. This reaction is similar to the Starch–Iodine Clock in Demonstration 47.

QUESTIONS FOR STUDENTS

1. What gas is produced in this reaction?
2. Propose a mechanism for this reaction.
3. Is this a redox reaction?
4. What might be done to renew the oscillation? Try it!

'4. An Oscillating Reaction: Red–Blue

everal solutions are placed in a petri dish. After about 5 min, the color of the solution oscillates between ɛd and blue.

rocedure

1. Prepare the oscillating solution as follows: Place 6 mL of solution *A* in a small beaker. Add 0.5 mL of solution *B*. Add 1.0 mL of solution *C*. A brown color will appear. When it disappears, add 1.0 mL of ferroin. Add 1 drop of *Photoflo* (or some other surface-active agent).
2. Add enough of this solution to a petri dish to half-fill it.
3. Wait for oscillations to begin.

:eactions

1. Bromate reacts with malonic acid to produce bromomalonate.
2. Bromate also reacts with red ferrous dye to produce blue ferric dye.
3. Bromide and malonic acid react to form bromomalonate.
4. Bromomalonate and blue ferric dye react to form bromide.
5. Bromide inhibits the reaction of red ferrous dye to blue dye, and a red color is produced.

olutions

1. Solution *A*: 5 g of sodium bromate and 2 mL of concentrated sulfuric acid in 67 mL of distilled water.
2. Solution *B*: 1 g of sodium bromide in 10 mL of distilled water.
3. Solution *C*: 1 g of malonic acid in 10 mL of distilled water.
4. Ferroin: 0.025 M solution.

'eaching Tips

NOTES

1. Ferroin is phenanthroline ferrous sulfate.
2. *Photoflo* can be found at any photography shop. It is a surface-active agent used in developing and printing.
3. This demonstration is adapted from a reaction described by Briggs and Raucher in "The Amateur Scientist," *Scientific American*, July, 1978.
4. This oscillating reaction does not require a magnetic stirrer!

QUESTIONS FOR STUDENTS

1. What is the role of ferroin in this reaction?
2. Can you propose a mechanism for this reaction?
3. Is this a redox reaction?
4. Are other ferrous salts red? Are ferric salts blue?

Oxidation-Reduction

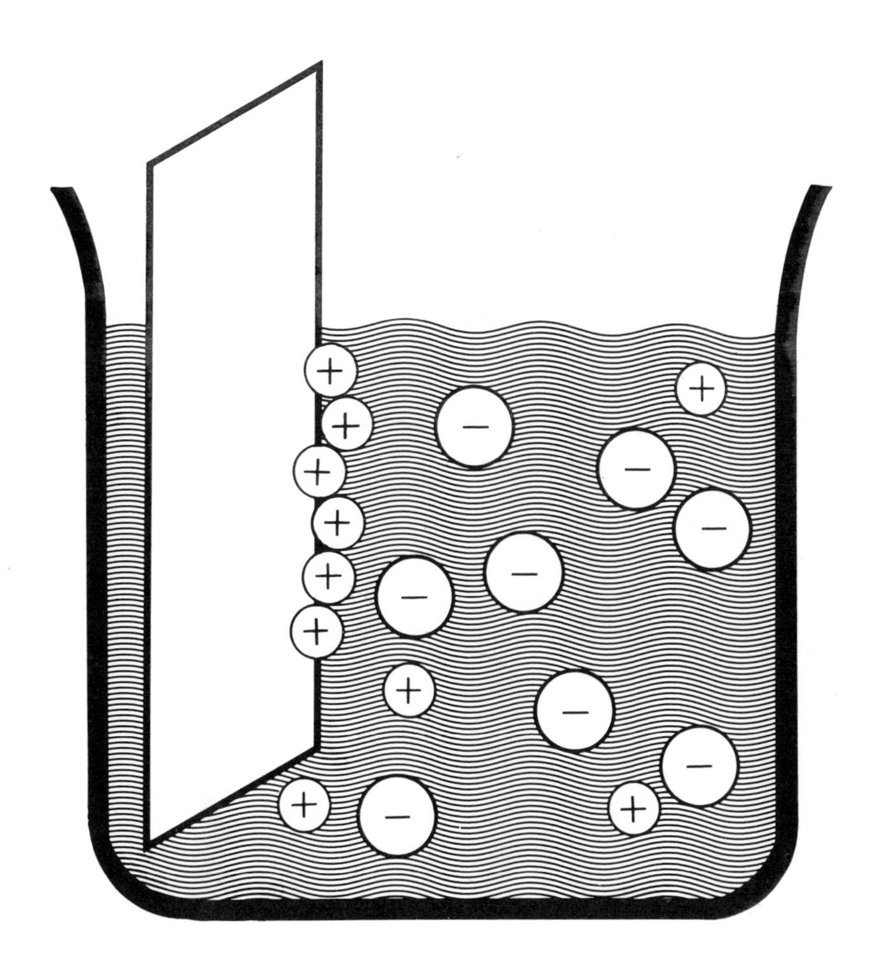

55. Catalytic Oxidation of Ammonia

A copper coil is heated in the flame of a bunsen burner until it is glowing hot. The coil is suspended above a layer of ammonium hydroxide in a flask. The coil continues to glow and eventually gets hot enough to melt.

Procedure

1. Prepare a copper coil ~ 1 in. in diameter by winding about 1 foot of bare copper wire (24 gauge works well) around a test tube. Form a hook on the end of the wire.
2. Place about 50 mL concentrated ammonium hydroxide in a 250-mL wide-mouth flask or beaker.
3. Heat the wire with a burner until it is glowing red-hot. CAREFUL!
4. Immediately place the wire in the flask, just above the level of the ammonium hydroxide.
5. Hook the end of the wire on the side of the flask.
6. Observe the reaction.
7. If the wire does not continue to glow, gently warm the ammonium hydroxide with a burner or add a small amount of sodium hydroxide solution.

Reaction

1. The hot copper causes ammonia to be oxidized with oxygen from the air:

$$4NH_3(g) + 3O_2(g) \xrightarrow{Cu} 2N_2(g) + 6H_2O(\ell) + \text{heat}$$

2. The reaction is highly exothermic and produces enough heat to melt the copper wire.

Teaching Tips

NOTES

1. Notice that bits of molten copper may fall into the solution. The blue color is due to the formation of copper complex ion.
2. You can also use a copper screen, which produces a more intense glow.
3. Wrap an old penny with copper wire. Heat the wire and penny and suspend it over methyl alcohol.
4. Darken the room for a more spectacular effect.

QUESTIONS FOR STUDENTS

1. What is the catalyst in this reaction?
2. What gas is produced in the reaction?
3. How hot does the reaction get?
4. Is this a redox reaction? What is oxidized? Reduced?

56. Oxidation of Glycerin by Permanganate

A small amount of crystalline substance is placed in a pile and a small amount of liquid added. After a few seconds a large puff of smoke and intense violet flames are produced.

Procedure

1. Place a small pile of granular potassium permanganate (about a tablespoonful) in the center of an evaporating dish.
2. Add 1 dropper full of glycerin on top of the pile.
3. STAND BACK! The reaction will occur in 15–20 s.

Reaction

$$14\,KMnO_4(g) + 4C_3H_5(OH)_3(\ell) \rightarrow 7K_2CO_3(s) + 7Mn_2O_3(s) + 5\,CO_2(g) + 16H_2O(\ell)$$

Solutions

1. Glycerin (glycerol) can be purchased at the drug store.
2. Use fresh, granular potassium permanganate ($KMnO_4$).

Teaching Tips

NOTES

1. Take special care with this reaction and all other reactions involving rapid oxidation.
2. You can tell when the reaction is about to begin—a slight puff of smoke will form in the center of the pile.
3. Keep in mind that this reaction is a *delayed* reaction!

QUESTIONS FOR STUDENTS

1. Is this a redox reaction?
2. If so, what is the oxidizer?
3. Does the size of the permanganate particles influence the rate of the reaction?
4. Write the equation for this reaction.
5. What is the gas produced?

57. Oxidation–Reduction: Iron

Colorless solutions, Fe^{2+}, become blood-red, $FeSCN^{2+}$, when oxidized. These red solutions again become colorless when reduced.

Procedure

OXIDATION

1. Label four beakers 1, 2, 3, and 4.
2. Place 50 mL solution *A* in each beaker.
3. Add 10 mL solution *B* to each beaker.
4. Add 10 mL of the following oxidizing agents to show formation of the red (III) complex: beaker 1, $K_2Cr_2O_7$; beaker 2, $KMnO_4$; and beaker 3, H_2O_2.

REDUCTION

Add $SnCl_2$ solution to any of the four beakers until the solution becomes colorless (forms II complex). Be patient; the reaction may take 30–60 s.

Reactions

OXIDATION

$$\underset{\text{(colorless)}}{Fe^{2+}(aq)} + \text{oxidizing agent} \rightarrow Fe^{3+}(aq)$$

$$Fe^{3+}(aq) + SCN^-(aq) \rightleftharpoons \underset{\text{(red)}}{FeSCN^{2+}(aq)}$$

1. Beaker 1: Dichromate ion, $Cr_2O_7^{2-}$, is reduced to Cr^{3+} in acid solution.
2. Beaker 2: Permanganate ion, MnO_4^-, is reduced to Mn^{2+} in acid solution.
3. Beaker 3: Peroxide is reduced to H_2O or OH^- in acid solution.

REDUCTION

$$\underset{\text{(red)}}{Fe^{3+}(aq)} + \underset{\text{(reducing agent)}}{Sn^{2+}(aq)} \rightarrow \underset{\text{(colorless)}}{Fe^{2+}(aq)} + Sn^{4+}(aq)$$

Solutions

1. Solution *A*: Dissolve 2 g of $Fe(NH_4)_2(SO_4)_2 \cdot 6H_2O$ (ferrous ammonium sulfate) in 20 mL of 6 M H_2SO_4 and dilute to 500 mL.
2. Solution *B* is 1 M KSCN: Dissolve 97 g of KSCN per liter.
3. The potassium dichromate, $K_2Cr_2O_7$, solution is 0.05 M: Dissolve 15 g of $K_2Cr_2O_7$ per liter.
4. The potassium permanganate, $KMnO_4$, solution is 0.05 M: Dissolve 7.5 g of $KMnO_4$ per liter.
5. The tin(II) chloride solution is 0.1 M: Dissolve 19 g of $SnCl_2$ per liter.
6. The hydrogen peroxide, H_2O_2, solution is 3% (drugstore variety).

Teaching Tips

NOTES

1. This demonstration shows the action of several oxidizing agents, as well as a reducing agent.
2. Cl_2 can also be used as an oxidizing agent. See Demonstration 7 for preparation of Cl_2 gas.
3. These reactions project well—use petri dishes and an overhead projector.
4. Solutions work best if prepared just before use.
5. See Demonstrations 89 and 92 for variations of this reaction.

QUESTIONS FOR STUDENTS

1. Write complete equations for the reaction produced in each demonstration.
2. How does Sn^{2+} act as a reducing agent?
3. Can you think of anything else that might act as an oxidizing agent for this reaction? Try it!
4. Can you detect any differences in the oxidizing power of these various oxidizing agents?

58. The Silver Mirror Reaction

Several solutions are placed in a round-bottom flask. The flask is swirled and a silver mirror forms to coat the inside of the flask.

Procedure

1. SCRUPULOUSLY clean a 50-mL Erlenmeyer flask.
2. Place 10 mL of solution *A* in the flask.
3. Mix 5 mL of solution *B* with 5 mL of solution *C* and add this mixture to the flask. (CAUTION!)
4. Quickly add 10 mL of solution *D*.
5. Stopper the flask and mix with a quick but gentle swirling motion. Be sure to evenly cover the entire glass surface with the solution while swirling.
6. Continue swirling until the silver mirror forms.
7. IMMEDIATELY pour the solution down the drain, and rinse the silvered flask with water.

Reaction

Metallic silver is formed when silver ion oxidizes the aldehyde part of a sugar molecule (glucose).

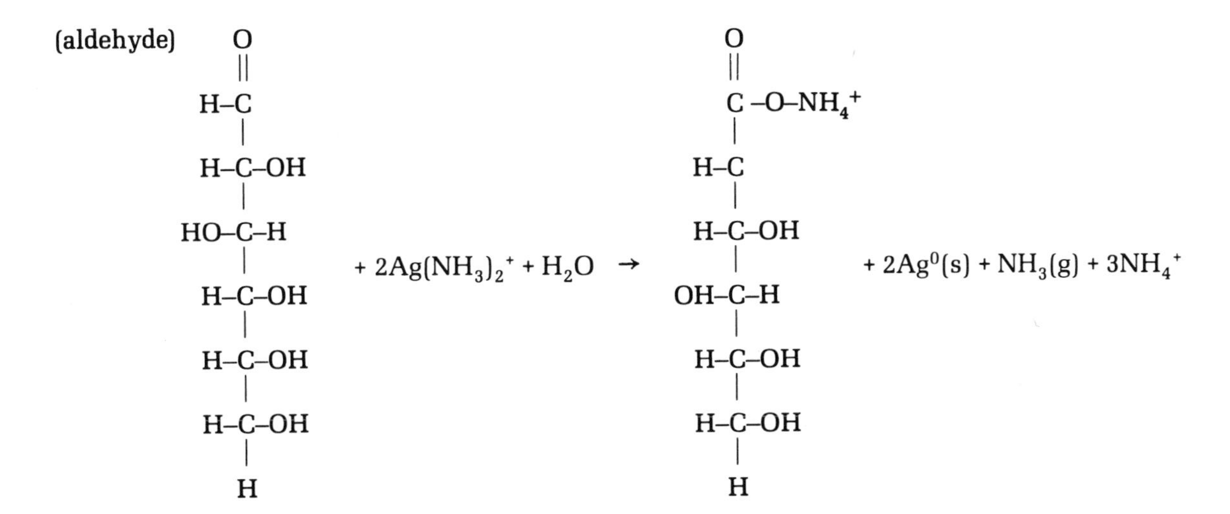

Solutions

1. Solution *A*: Dissolve 2.5 g of glucose and 2.5 g of fructose in 50 mL of water. Add 0.6 g of tartaric acid. Bring the solution to a boil, then cool. Add 10 mL of ethyl alcohol and dilute to 100 mL.
2. Solution *B*: Dissolve 4.0 g of silver nitrate in 50 mL of distilled water.
3. Solution *C*: Dissolve 6.0 g of ammonium nitrate in 50 mL of water.
4. Solution *D*: Dissolve 10 g of sodium hydroxide in 100 mL of water.

Teaching Tips

NOTES

1. This demonstration effectively shows the reduction of silver ion to silver metal and

2. If the mirror does not form, the flask was probably not clean.
3. The ammoniacal silver solution produced by this reaction is potentially explosive! Therefore, the following precautions MUST be taken:

   a. Do not mix solution *B* and solution *C* until the demonstration is performed.
   b. When the silver mirror forms, IMMEDIATELY flush the reaction solution down the drain and rinse the silvered flask with water.

4. If desired, remove the mirror by dissolving in concentrated HNO_3 (CAREFUL!).

QUESTIONS FOR STUDENTS

1. Write a chemical equation for the reaction.
2. What is the silver mirror?
3. Do you think that this method is a good way to produce a mirror? Is it economically feasible?
4. What is the role of the sugar in this reaction?
5. Which hydrogen was removed from the sugar molecule when it was oxidized? (The hydrogen on the aldehyde group was removed.)

59. The Blue Bottle Reaction

A large round-bottom flask contains a colorless solution. When shaken vigorously, a blue color forms. After a few seconds, the blue color fades and the solution again becomes colorless. The process can be repeated.

Procedure

1. Show your students a stoppered flask containing ∿ 300 mL of solution.
2. Give the flask a few vigorous shakes. Notice that the solution turns blue.
3. After a few seconds, the solution becomes clear again.
4. Ask for suggestions to explain these reactions.
5. The process can be repeated several times. It may be necessary to periodically remove the stopper. (Why?)

Reactions

1. The students should deduce that the reaction causing the blue color is essentially

   $$\text{gas} + \text{liquid} \rightarrow \text{blue color}$$

   and that the clearing of the solution is simply

   $$\text{blue color} + x \rightarrow \text{colorless}$$

2. The actual reaction involves the reduction of methylene blue by an alkaline dextrose solution. Upon shaking, the reduced product is reoxidized to the blue dye. Essentially, the reaction occurs in four steps:

   $$O_2(g) \rightarrow O_2 \text{ (dissolved)}$$

   $$O_2 \text{ (dissolved)} + \underset{\text{(colorless)}}{\text{methylene blue}} \rightarrow \underset{\text{(blue)}}{\text{methylene blue}}$$

   $$\text{glucose} + OH^- \rightarrow \text{glucoside}$$

   $$\text{glucoside} + \underset{\text{(blue)}}{\text{methylene blue}} \rightarrow \underset{\text{(colorless)}}{\text{methylene blue}} + OH^-$$

Solutions

Prepare the solution for the flask as follows: Add 8 g of KOH to 300 mL of water. Cool and add 10 g of glucose (dextrose). Add a few drops of methylene blue indicator solution, or a small (matchhead size) amount of the solid indicator. DO NOT ADD TOO MUCH!

Teaching Tips

NOTES

1. This reaction is an excellent inquiry demonstration. Challenge your students to suggest why the blue color is produced and why it later disappears.

2. Students should be expected to deduce the overall reaction, but not the formation of the glucoside or the hydroxide.
3. The solution is good for only about 12–15 shakings.

QUESTIONS FOR STUDENTS

1. How do you know that a reaction occurred?
2. Could something on the stopper cause the reaction? (Remove the stopper and shake the flask—the blue color is still produced.)
3. What do you notice at the gas–liquid interface?
4. Could this interface be where the reaction is occurring? (YES!)
5. Is the *debluing* simply a reverse of the *bluing*?
   (No; if it were, gas would be evolved. Also, a reverse reaction doesn't occur unless conditions are changed. Conditions are the same!)
6. What does the shaking do? (It dissolves oxygen.)
7. Why does the reaction eventually *run down*?

60. Oxidation of Zinc: Fire and Smoke

A drop of water is added to a small pile of chemicals in an evaporating dish. After a few seconds, an intense blue flame with smoke is produced.

Procedure

1. Mix (DO NOT GRIND) ammonium nitrate and ammonium chloride in a 4:1 ratio to give enough mixture to half fill a small evaporating dish.
2. Sprinkle enough zinc dust on the mixture to lightly cover its surface.
3. Add about 3 drops of water from a dropper, letting the water run down the side of the dish.
4. STAND BACK!

Reactions

1. Cl^- (from NH_4Cl) acts as a catalyst on the decomposition of NH_4NO_3:

$$NH_4NO_3(s) \xrightarrow{Cl^-} N_2O(g) + 2H_2O(aq)$$

2. Water produced in the reaction causes the decomposition of more NH_4NO_3 (autocatalytic effect).
3. The reaction melts the NH_4NO_3 and allows the oxidation of the zinc. The overall reaction is probably as follows:

$$Zn(s) + NH_4NO_3(s) \rightarrow N_2(g) + ZnO(s) + 2H_2O(g)$$

Teaching Tips

NOTES

1. The chemicals in this reaction represent a mixture of oxidizing and reducing agents. Water acts as a catalyst.
2. This reaction is highly exothermic. TAKE NECESSARY PRECAUTIONS!
3. This reaction produces a dense cloud of white ZnO (s). You can add a few crystals of iodine to produce purple smoke. This reaction must be done in a hood or a well-ventilated area.
4. The oxidizing properties of nitrates are demonstrated.

QUESTIONS FOR STUDENTS

1. Write the chemical equation for the reaction.
2. What is the catalyst? What is *autocatalysis*?
3. What was oxidized? What was reduced?
4. What is the *smoke*?
5. What was the cause of the short delay before this reaction began?

61. The Mercury Beating Heart

A small amount of mercury is placed in a solution in a watch glass. When touched with a needle, the mercury pulsates in a rhythmic fashion, resembling a beating heart.

Procedure

1. Place a large watch glass inside a petri dish.
2. Add CLEAN mercury to the watch glass to form a pool no more than three-fourths of an inch in diameter.
3. Add dilute sulfuric acid until the surface of the mercury pool is just covered.
4. Drop 1 mL of potassium dichromate solution on top of the mercury.
5. Place a clean needle on the watch glass so that it just barely touches the mercury pool. Firmly fix the needle in position with putty.
6. The mercury will soon begin to pulsate.
7. To get a stronger beat, slowly add concentrated sulfuric acid, drop by drop.

Reactions

1. The Hg pool forms a sphere because of a large electrical charge (i.e., number of electrons) on the surface.
2. $Cr_2O_7^{2-}$ acts as an oxidizing agent. Electrons are removed from the Hg, and the Hg drop flattens.
3. When the drop flattens, it touches the nail or needle and receives electrons.

$$Cr_2O_7^{2-}(aq) + 14H^+(aq) + 6e^- \rightarrow 2Cr^{3+}(aq) + 7H_2O(\ell)$$

$$Fe(aq) \rightarrow Fe^{3+}(aq) + 3e^-$$

4. The increased number of electrons on the Hg causes it to again become spherical, and it moves away from the needle.
5. The process is repeated.

Solutions

1. The dilute sulfuric acid is 1.0 M: 1 mL of concentrated H_2SO_4 in 17 mL of water.
2. Enough concentrated H_2SO_4 to strengthen the beat.
3. The potassium dichromate solution is 0.1 M: 2.9 g of $K_2Cr_2O_7$ per 100 mL.

Teaching Tips

NOTES

1. Reagent-grade Hg works best. It must be pure and clean.
2. The watch glass is placed in a petri dish so that it can be moved easily and to catch any accidental spillage.
3. Put the dish on the overhead projector!
4. You will see several patterns of oscillations (beats): a small to large sphere, an equilateral triangle (most common type), or a four-lobed shape.
5. A very thin iron wire works well.

QUESTIONS FOR STUDENTS

1. Explain what is happening when the mercury pool contracts and expands.
2. Why is mercury an ideal substance to use in this demonstration?
3. Is a catalyst used in this reaction?
4. Does the mercury settle into a regular beating pattern? Why?

62. Oxidation States of Manganese: Quick Mn^{6}

A solution of purple Mn^{7+} is poured over a folded white paper napkin and is immediately reduced to the green Mn^{6+}.

Procedure

1. Place 20 mL of potassium permanganate solution in a small beaker.
2. Add a few milliliters (the amount is not critical) of sodium hydroxide solution.
3. Mix the two solutions. Note the purple color that is characteristic of the Mn^{7+} ion.
4. Fold a heavy white paper napkin several times and pour the solution over the napkin.
5. Note the immediate and intense green color of the Mn^{6+} ion on the napkin.

Reaction

The cellulose in the napkin reduces the alkaline Mn^{7+} to Mn^{6+}.

Solutions

1. The potassium permanganate is 0.01 M: Dissolve 1.6 g of $KMnO_4$ per liter.
2. The sodium hydroxide solution is 2 M: Dissolve 80 g of NaOH per liter.

Teaching Tips

NOTES

1. See Demonstration 84 for ways to show other oxidation states of manganese.
2. Get white napkins from the school cafeteria. Filter paper will work well also.
3. This reaction is similar to that shown in Demonstration 59. The cellulose in paper is a polymer of dextrose molecules.

QUESTIONS FOR STUDENTS

1. Will this reaction occur if the solution is poured over a white handkerchief instead of a napkin? Try it!
2. What does the napkin do?
3. Write an equation to show this reaction.
4. Does the green color persist?

63. Oxidation States of Manganese: Mn^{7+}, Mn^{6+}, Mn^{4+}, and Mn^{2+}

Three reactions are carried out with potassium permanganate, producing colors characteristic of the various oxidation states of manganese.

Procedure

1. Label four large beakers *A*, *B*, *C*, and *D*.
2. Place 50 mL of $KMnO_4$ solution in each beaker.
3. Set beaker *D* aside; it represents manganese in the +7 oxidation state.
4. Add 10–15 mL of H_2SO_4 to beaker *A*.
5. Add 20 mL of NaOH to beaker C.
6. Place beaker *A* on a white background. Slowly add $NaHSO_3$ solution while stirring. Note the change of color to red, pink, and finally colorless. This colorless solution indicates the +2 oxidation state.
7. Place beaker *B* on the same background and add $NaHSO_3$ while stirring. Note the formation of a brown precipitate, indicating the +4 state.
8. Place beaker C on the same background and add $NaHSO_3$ while stirring. Note the formation of a green color, indicating the +6 state.
9. Arrange the four beakers to show the colors and the +7, +6, +4, and +2 oxidation states.

Reactions

1. In beaker *A* (+7 to +2)

$$2MnO_4^-(aq) + H^+(aq) + 5HSO_3^-(aq) \longrightarrow 2Mn^{2+}(aq) + 5SO_4^{2-}(aq) + 3H_2O(\ell)$$

(purple) (colorless)

2. In beaker *B* (+7 to +4)

$$OH^- + 2MnO_4^-(aq) + 3HSO_3^-(aq) \longrightarrow 2MnO_2(s) + 3SO_4^{2-}(aq) + 2H_2O(\ell)$$

(purple) (brown precipitate)

3. In beaker C (+7 to +6)

$$2MnO_4^-(aq) + 3OH^-(aq) + HSO_3^-(aq) \longrightarrow 2MnO_4^{2-}(aq) + SO_4^{2-}(aq) + 2H_2O(\ell)$$

(purple) (green)

Solutions

1. The potassium permanganate solution is 0.01 M: Dissolve 0.74 g of $KMnO_4$ per 500 mL.
2. The H_2SO_4 solution is 3 M (*see* Appendix 2).
3. The sodium hydroxide solution is 2.0 M: dissolve 8 g of NaOH per 100 mL.
4. The sodium bisulfite solution is 0.01 M: Dissolve 0.54 g of $NaHSO_3$ per 500 mL.

Teaching Tips

NOTES

1. Be sure to use sodium bisul*fite*, not sodium bisul*fate*.
2. MnO_4^-(+7) is purple; MnO_4^{2-} (+6) is green; MnO_2 (+4) is a brown precipitate; and Mn^{2+} is colorless to light pink.

QUESTIONS FOR STUDENTS

1. Write the chemical equation for these reactions.
2. List the different oxidation states of manganese and their colors.
3. Does manganese have oxidation states not demonstrated here? (Yes. Violet +3 ion as in Mn_2O_3 and $MnCl_3$.)
4. Are some oxidation states easier to obtain than others?
5. Indicate what was oxidized and what was reduced in each reaction.

34. The *Prussian Blue* Reaction

'wo colorless solutions are mixed. After a few seconds the mixture turns yellow, then green, and finally deep blue (Prussian blue).

Procedure

1. Mix equal volumes of solutions *A* and *B* in a large beaker.
2. Observe the color changes.

Reactions

1. This reaction involves a series of oxidation–reduction reactions. The ferric ion reacts with the ferrocyanide ion to produce a blue complex called Prussian blue.
2. The ferric ion is almost completely reduced to ferrous ion by ferrocyanide ion, which is simultaneously oxidized to ferricyanide ion.
3. The first reaction produces the yellow hexacyanoferrate(II) ion, $Fe(CN)_6^{4-}$.
4. The second reaction involves the mixing of the yellow ion with the blue $Fe(III)[Fe(II)(CN)_6]$ species to produce a green color.
5. The final reaction is the production of the blue complex, $KFe[Fe(CN)_6]$.

$$Fe^{2+}(aq) + 2CN^-(aq) \rightarrow Fe(CN)_2(s)$$

$$Fe(CN)_2(s) + 4CN^-(aq) \rightarrow \underset{\text{hexacyanoferrate(II) (yellow)}}{[Fe(CN)_6]^{4-}(aq)}$$

$$K^+(aq) + [Fe(CN)_6]^{4-}(aq) + Fe^{3+}(aq) \rightarrow \underset{\text{(Prussian blue)}}{KFe[Fe(CN)_6]}$$

Solutions

1. Solution *A* is 0.0002 M $K_4Fe(CN)_6 \cdot 3H_2O$. This solution is obviously very dilute! Prepare it by dissolving 0.422 g of potassium ferrocyanide in 100 mL of water to make a 0.01 M solution. Dilute this solution to 0.0002 M by diluting 1 mL of solution to 50 mL.
2. Solution *B* is 0.0002 M $NH_4Fe(SO_4)_2 \cdot 12H_2O$. NOTE: THIS SOLUTION IS NOT IN WATER; IT IS A SOLUTION IN POTASSIUM BISULFATE. First prepare a 0.01 M solution by dissolving 0.48 g of ferric ammonium sulfate in 100 mL of 0.1 M $KHSO_4$ solution. Dilute 1 mL of this solution in 50 mL of the $KHSO_4$ solution to prepare the 0.0002 M solution.
3. $KHSO_4$ solution is 0.1 M. Dissolve 2.72 g of $KHSO_4$ in 200 mL of water.

Teaching Tips

NOTES

1. Proper concentration is essential for formation of all three colors. Keep trying until you get distinct and sharp color changes.
2. This method is a colorful way to show a reaction in a series of steps before the final oxidation–reduction product is obtained.

3. This reaction is responsible for the production of *blueprints*. See Demonstration 72 for this application of the Prussian blue reaction.
4. Prussian blue has the formula $KFe[Fe(CN)_6]$.

QUESTIONS FOR STUDENTS

1. Write the chemical equations for the reactions.
2. Which of the observed reactions is indicative of the rate of the overall reaction?
3. Are some reaction rates different? Why?
4. Determine the oxidizing and reducing agents.

65. The Activity Series for Some Metals

A petri dish is placed on an overhead projector. HCl is added to the dish and five metal pieces are placed in labeled areas. After a few seconds, bubbles of hydrogen gas appear on the metals. The hydrogen bubbles form at a rate relative to the activity of each metal.

Procedure

1. Outline the circumference of a glass petri dish with a felt-tip marker on a plastic sheet. Label the areas on the sheet Cu, Fe, Sn, Zn, and Mg.
2. Place the plastic sheet on an overhead projector and place the dish over the drawn circle. The labeled symbols should project as if within the dish.
3. Cover the bottom of the dish with 6 M HCl.
4. Carefully place small pieces of Cu, Fe, Sn, Zn, and Mg near their respective labels.
5. Note the extent of reaction of each metal with HCl.

Reaction

The general reaction for a metal with an acid is

$$\text{metal(s)} + 2HCl(aq) \rightarrow H_2(g) + \text{metal } Cl^-(aq)$$

Solutions

The HCl solution is 6 M (see Appendix 2).

Teaching Tips

NOTES

1. This demonstration is an excellent way to simultaneously view five reactions.
2. The order, according to rate of hydrogen gas formation, is Mg > Zn > Fe > Sn > Cu.

QUESTIONS FOR STUDENTS

1. Write the chemical equation for each reaction.
2. What observed reaction is indicative of the rate of the reaction?
3. Which metal reacts best with HCl? Which metal reacts least?
4. Would a different acid change the activity order? Try it!
5. Would a different acid concentration change the activity order? Try it!

66. Copper into Gold: The Alchemist's Dream!

A copper penny is placed in an evaporating dish and heated with a mixture. It turns *silver*. The penny is then heated in a burner flame and it suddenly turns *gold*!

Procedure

1. Place ∿ 5 g of zinc dust in an evaporating dish.
2. Add enough NaOH solution to cover the zinc and fill the dish about one-third.
3. Place the dish over a burner, or hot plate, and heat until the solution is near boiling.
4. Prepare a copper penny by cleaning it thoroughly with a light abrasive (Brillo pads work well).
5. Using crucible tongs, or tweezers, place the cleaned penny in the mixture in the dish.
6. Hold the penny in the dish for 3–4 minutes. You will be able to tell when the coating is complete.
7. Remove the penny, wash it, and blot dry with paper towels. (Do not rub.)
8. Using tweezers, hold the coated penny in the flame of a burner. The production of the *gold* color is immediate.
9. After 3–5 s, remove the coin, wash it, and dry it.

Reactions

1. The first reaction is the plating of the copper with zinc: Zinc reacts with sodium hydroxide to form sodium zincate, Na_2ZnO_2. This product is then reduced by the copper penny to metallic zinc. This reaction gives the silver color to the penny.
2. The second reaction is the formation of the brass alloy. This alloy gives the penny the gold color. Heat causes a fusion of the zinc and copper.

Solution

The sodium hydroxide solution is 6 M: Dissolve 240 g of NaOH per liter.

Teaching Tips

NOTES

1. You may not want to deny your students the opportunity to perform this demonstration as an experiment—they love it!
2. New pennies seem to work best, if not overheated. Pre-1980 pennies are copper; later pennies have a zinc core.
3. This demonstration can be used to illustrate a variety of reactions: solid–solid reactions, oxidation–reduction reactions, or metallurgy.
4. Brass is 60–82% Cu and 18–40% Zn.

QUESTIONS FOR STUDENTS

1. Is this reaction an oxidation–reduction reaction?
2. Why did the penny turn silver?
3. Why did it turn "gold"?
4. Why did the penny need to be heated in order to turn "gold"?

67. Oxidation of Sodium

A small piece of sodium metal is dropped into a water-filled petri dish on an overhead projector. The sodium forms a small ball and darts around the dish, leaving a pink trail.

Procedure

1. Place the smaller half of a petri dish on an overhead projector.
2. Half fill the dish with water.
3. Add 1 drop of phenolphthalein and 1 grain of laundry detergent.
4. Stir the contents of the dish thoroughly.
5. Using tweezers, place a VERY SMALL (pinhead size) piece of freshly cut sodium metal in the dish. (CAUTION!)
6. Cover the dish with the other half of the petri dish.
7. Observe the random motion of the sodium and the color change in the solution.

Reactions

1. The sodium is oxidized:

$$\text{Na metal (s)} \rightarrow Na^{+}(aq) + e^{-}$$

2. The water is reduced:

$$2H_2O(\ell) + 2e^{-} \rightarrow H_2(g) + 2OH^{-}(aq)$$

3. The basic solution reacts with the phenolphthalein to produce the pink trail behind the sodium.

Teaching Tips

NOTES

1. This reaction is an excellent way to demonstrate acid–base reactions, oxidation–reduction reactions, the action of an indicator, and the activity of metallic sodium.
2. DO NOT USE larger pieces of sodium. AN EXPLOSION CAN OCCUR when large pieces of sodium are placed in water.
3. The dish is covered to prevent spattering on the projector glass.
4. If the sodium gets stuck on the side of the dish, give it a push with the tweezers to keep it moving.

QUESTIONS FOR STUDENTS

1. Why was detergent added?
2. How does the sodium behave?
3. How can you explain this behavior on a chemical basis?
4. What causes the pink trail?

68. Oxidation of Alcohol by Mn_2O_7

A large test tube is mounted on a ringstand and immersed in a large beaker of water. Several crystals are dropped into the tube. Bubbles of gas appear at the interface of the two liquids in the tube, a green color is produced, and tiny sparks are produced that pop and flash.

Procedure

1. Secure a large test tube to a ringstand by attaching the clamp at the top of the tube.
2. Slowly add concentrated sulfuric acid until the tube is one-fourth full.
3. With a pipet, add the same amount of ethyl alcohol. Pour it slowly down the side of the tube, so that the two liquid layers do not mix.
4. Immerse the tube in a large beaker of water, so that at least half of the acid layer is under water.
5. Add a few crystals of potassium permanganate to the tube. You may need to grind the $KMnO_4$ with a mortar and pestle.
6. Observe the reaction that will occur in a couple of minutes.

Reactions

1. Green Mn_2O_7 is formed at the interface of the two liquids.
2. Mn_2O_7 is a powerful oxidizing agent that oxidizes the alcohol at the acid–alcohol interface.

$$KMnO_4(s) + 3H_2SO_4(aq) \rightarrow K^+(aq) + MnO_3^+(aq) + H_3O^+(aq) + 3HSO_4^-(aq)$$

$$MnO_3^+(aq) + MnO_4^-(aq) \rightarrow Mn_2O_7(s)$$

$$Mn_2O_7(s) + CH_3CH_2OH(aq) \rightarrow 2CO_2(g) + 3H_2O(\ell) + 4MnO_2(aq)$$

Solutions

1. The sulfuric acid is concentrated.
2. The ethyl alcohol is about 95%.

Teaching Tips

NOTES

1. This demonstration is beautiful, but POTENTIALLY DANGEROUS! Use extreme caution with the acid.
2. The large beaker of water is an added precaution.
3. Do not try to stop this reaction; let it run to completion. First, use a pipet to remove the alcohol layer. Next, slowly dilute the acid by running water down the side of the tube, then discard the solution.
4. Try adding only a few crystals at first. As the reaction slows, you can add more.
5. Mn_2O_7 is the anhydride of permanganic acid. In concentrated form, it is a dark brown, highly explosive liquid.
6. You may also notice some accumulation of MnO_2; it is also one of the products of reduction.

QUESTIONS FOR STUDENTS

1. Why is sulfuric acid added? (Manganese is a stronger oxidizing agent in acid than it is in base.)
2. Trace the oxidation state of manganese through the series of reactions.
3. What is the green compound?
4. Why does it *flash*?

69. Oxidation States of Vanadium: Reduction of V^{5+} to V^{2+}

A flask containing a yellow solution is gently shaken. The solution turns blue. The flask is shaken again and the solution turns green. More vigorous shaking produces a violet color.

Procedure

1. Place 50 g of zinc amalgam in a 1-L Florence flask.
2. Add 200 mL of vanadium solution.
3. Stopper the flask.
4. Note the color of the solution. Pour a little of the solution in a beaker as a sample.
5. Gently swirl the flask and note the blue color. Save a small amount as a sample.
6. Shake the flask and note the green color. Save a sample.
7. Shake the flask more vigorously and note the violet color. Save a sample.

Reactions

REACTION 1

$$Zn(s) + VO_3^-(aq) + 8H_3O^+(aq) \rightleftharpoons VO^{2+}(aq) + Zn^{2+}(aq) + 12H_2O(\ell)$$

REACTION 2

$$Zn(s) + VO^{2+}(aq) + 4H_3O^+(aq) \rightleftharpoons 2V^{3+}(aq) + Zn^{2+}(aq) + 6H_2O(\ell)$$

REACTION 3

$$Zn(s) + 2V^{3+}(aq) \rightleftharpoons 2V^{2+}(aq) + Zn^{2+}(aq)$$

SUMMARY OF REACTIONS

$$VO_3^-(aq) \rightarrow VO^{2+}(aq) \text{ (yellow to green)}$$

$$VO^{2+}(aq) \rightarrow V^{3+}(aq) \text{ (green to blue)}$$

$$V^{3+}(aq) \rightarrow V^{2+}(aq) \text{ (blue to violet)}$$

Solutions

1. Zinc amalgam: In a 500-mL flask, dissolve 2 g of mercuric chloride in 300 mL of water. Add 2 mL of concentrated nitric acid. Add 250 g of zinc (20–30 mesh works best). Stopper the flask and shake it for a few minutes. Pour off the liquid and wash the amalgam several times with water. Store the amalgam under water in a sealed jar.
2. Vanadium solution: In a large beaker, dissolve 6–8 g of NaOH in 200 mL of water. Add 10 g of ammonium vanadate (NH_4VO_3). You may need to warm the beaker to completely dissolve this compound. Stir constantly. Add 500 mL of sulfuric acid solution (prepared by adding 50 mL of concentrated sulfuric acid to 450 mL of water). Dilute to 1 L. Store in a large, well-stoppered bottle.

Teaching Tips

NOTES

1. These colors project well and can be used in petri dishes on an overhead projector.
2. The used amalgam can be washed several times and reused.
3. Vanadium is named for *Vanadis*, the Norse Goddess of Beauty. (Why?)

QUESTIONS FOR STUDENTS

1. What is an amalgam?
2. Write the equation for the reaction producing each colored vanadium ion.
3. What is oxidized and what is reduced in each reaction?
4. What does shaking do?

70. Oxidation States of Vanadium: Reoxidation of V^{2+} to V^{5+}

Demonstration 69 reduced vanadium through the +5, +4, +3, and +2 oxidation states. This demonstration reverses the process by reoxidizing the vanadium solution through the +2, +3, +4, and, finally, +5 states.

Procedure

1. Fill a buret with ceric sulfate solution.
2. Add this solution, drop by drop, to the violet solution from Demonstration 69, swirling constantly.
3. Note the appearance of colors characteristic of the various oxidation states as more ceric sulfate solution is added.

Reactions

REACTION 1

$$Ce^{4+}(aq) + V^{2+}(aq) \rightleftharpoons V^{3+}(aq) + Ce^{3+}(aq)$$

REACTION 2

$$Ce^{4+}(aq) + V^{3+}(aq) + 3H_2O(\ell) \rightleftharpoons VO^{2+}(aq) + Ce^{3+}(aq) + 2H_3O^{+}(aq)$$

REACTION 3

$$Ce^{4+}(aq) + VO^{2+}(aq) + 6H_2O(\ell) \rightleftharpoons VO_3^{-}(aq) + Ce^{3+}(aq) + 4H_3O^{+}(aq)$$

SUMMARY

$$V^{2+}(aq) \rightarrow VO^{3+}(aq) \text{ (violet to blue)}$$

$$V^{3+}(aq) \rightarrow VO^{2+}(aq) \text{ (blue to green)}$$

$$VO^{2+}(aq) \rightarrow VO_3^{-}(aq) \text{ (green to yellow)}$$

Solutions

1. Vanadium solution: Use the solution prepared for Demonstration 69.
2. The ceric sulfate solution is 1.0 M: Add 33.2 g of $Ce(SO_4)_2$ to 100 mL of 1.0 M H_2SO_4 (see Appendix 2).

Teaching Tips

NOTES

1. Sulfuric acid is added to the ceric sulfate solution to increase the solubility of the salt.
2. Demonstrations 69 and 70 could be done together, but a lot of chemistry is involved!

QUESTIONS FOR STUDENTS

1. Write the reactions for the reoxidation of vanadium.
2. What color is charactistic of each oxidation state?
3. To what group of elements does vanadium belong?

71. Oxidation States of Chromium

Students are shown colored solutions of various chromium ions. One solution is green; another is red-orange. Hydrogen peroxide is added to the red-orange solution and a dark blue solution is produced.

Procedure

1. Shake a little chromic oxide (Cr_2O_3) in a container with water. The green color is characteristic of the Cr^{3+} ion.
2. Display a solution of chromic(VI) oxide, CrO_3. This solution is red-orange, characteristic of the Cr^{6+} ion.
3. Pour a small amount of CrO_3 solution in another beaker and add a few milliliters of hydrogen peroxide.
4. Note the color change to blue, characteristic of the Cr^{2+} ion.

Reactions

Cr^{6+} is reduced by peroxide to form Cr^{2+}.

$$Cr^{6+}(aq) + H_2O_2(aq) \rightarrow Cr^{2+}(aq) + 2H_2O(\ell) + O_2(g)$$

(red-orange) (blue)

Solutions

1. All solutions should be ∿ 1.0% (10 g per liter)
2. The hydrogen peroxide is 3% H_2O_2 (from the drug store).

Teaching Tips

NOTES

1. CrO_3 is a strong oxidizing agent.
2. The most *stable* oxidation state of chromium is +3.
3. Cr^{2+} is easily air oxidized to Cr^{3+}.
4. Chromic(VI) oxide, CrO_3, forms dichromic acid ($2H^+ + Cr_2O_7^{2-}$) in aqueous solution. This is an equilibrium mixture of

$$Cr_2O_7^{2-}(aq) + H_2O(\ell) \rightarrow 2CrO_4^{2-}(aq) + 2H^+(aq)$$

5. A summary of the oxidation states of chromium is as follows:

Cr^{2+}, as in $CrCl_2$, is blue; Cr^{3+}, as in Cr_2O_3, is green; and Cr^{6+}, as in CrO_3, is red-orange.

QUESTIONS FOR STUDENTS

1. Which oxidation state is associated with chromium in potassium dichromate? Potassium chromate?
2. Write the reaction for the action of hydrogen peroxide on CrO_3.
3. Give the color characteristic of each chromium ion.

72. Photoreduction: The *Blueprint* Reaction

key, or some similar object, is placed on a sheet of blue-green paper. After exposing this paper to a loodlight for a few minutes, the paper is washed and an imprint of the key is left on the paper.

Procedure

1. Presoak a light-sensitive paper by coating ordinary bond paper with the photosensitive solution (see Solutions).
2. Allow the treated paper to dry in a dimly lit room.
3. Place an object such as a key, comb, or pencil on the paper.
4. Expose the paper to a strong light (floodlight or outside sunlight) for 2–3 min.
5. Wash the exposed paper in running water.
6. Note the imprint.

Reactions

1. The photosensitive compound, iron(III)hexacyanoferrate(III), $Fe[Fe(CN)_6]$, is formed when the two solutions are mixed.
2. $Fe[Fe(CN)_6]$ is reduced by light to form iron(III) hexacyanoferrate(II), $Fe_4[Fe(CN)_6]_3$.
3. The overall reaction is

   A. Coating the paper:

$$Fe^{3+}(aq) + Fe(CN)_6^{3-} \rightleftharpoons \underset{\text{(bronze-green)}}{Fe[Fe(CN)_6](aq)}$$

   B. Photoreduction on the paper:

   a. $Fe^{3+} \xrightarrow[\text{citrate}]{\text{light}} Fe^{2+}$

   b. $3Fe^{2+} + Fe(CN)_6^{3-} \rightarrow \underset{\text{(blue)}}{Fe_3[Fe(CN)_6]_2}$

Solutions

1. Make two solutions:

   Solution *A*: Dissolve 30 g of potassium hexacyanoferrate(III), $K_3Fe(CN)_6$, in 100 mL of water.
   Solution *B*: Dissolve 40 g of iron(III) ammonium citrate in 100 mL of water.

2. Mix these two solutions in a dimly lit room (no direct sunlight) to form the photosensitive solution.

Teaching Tips

NOTES

1. You can easily coat the paper by using a sponge soaked in solution. Use a paper that does not soak up the solution.
2. For best results, keep the dry treated paper in a drawer or box prior to its use.

3. Wash your hands immediately after treating the paper.
4. Place a black and white negative on the paper, cover both with a glass plate, and expose to strong light for about 5 min. When the paper is washed, a crude photograph will be produced.

QUESTIONS FOR STUDENTS

1. Write the equation for the blueprint reaction.
2. What role does light play in this reaction?
3. What is the oxidation state of iron in each of the complexes formed in this reaction?
4. Can you give other examples of photoreduction reactions? (Silver halides are reduced by light when a picture is taken.)

Electrochemistry

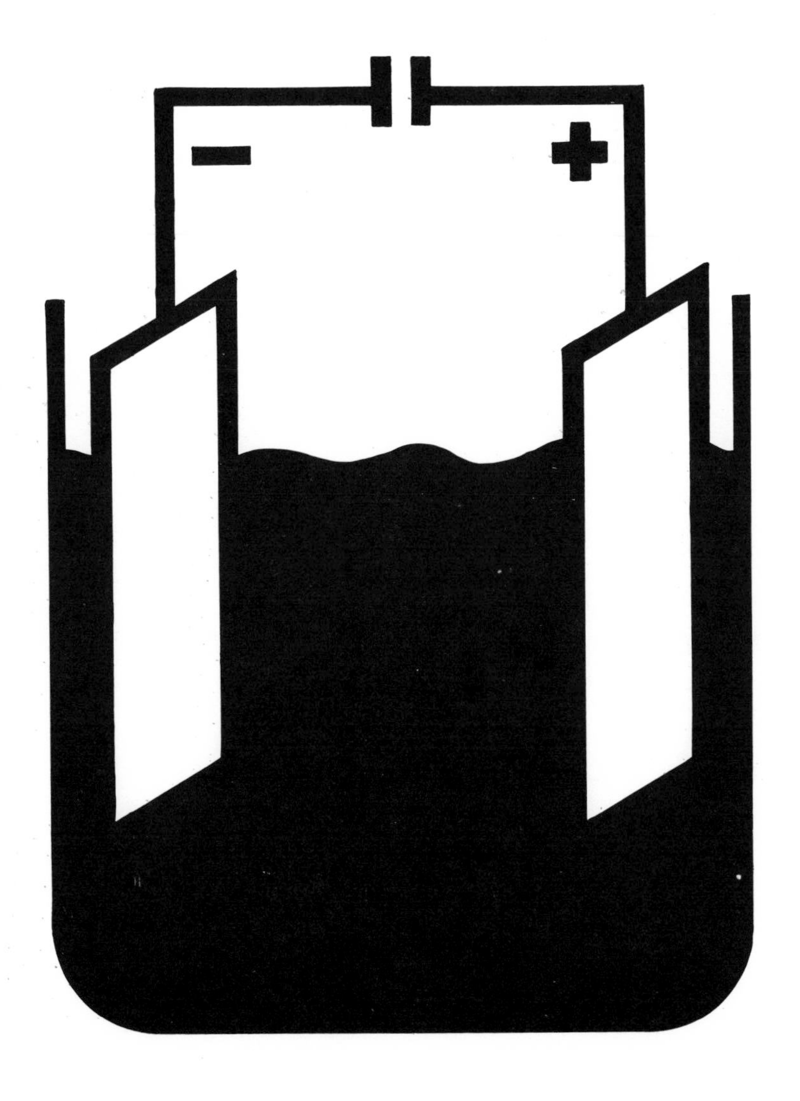

73. Making a Simple Battery: The *Gerber* Cell

A simple battery is made from a large baby food jar. When completed, the battery will generate 1.5 V and light a small bulb. Six of these batteries connected in series will operate a small 9-V pocket radio.

Procedure

1. Construct the battery according to Figure 3. Use a large baby food jar and a one-hole rubber stopper (#9). Beakers can be used also.
2. Cut a strip of copper metal long enough to fit into the jar; allow approximately 1 in. extra. Cut a magnesium strip the same length.
3. If the copper strip is not shiny, dip it into a dilute HNO_3 solution for a few seconds. Wash thoroughly in running water.
4. Cut a 6-in. length of dialysis tubing and hold it under water until it becomes flexible. Tie a knot in one end to make a bag. Insert the copper strip in the dialysis bag and fill the bag with copper sulfate solution. Place the prepared bag in the jar.
5. Place the magnesium strip in the jar and fill the jar with sodium sulfate solution.
6. Insert the stopper so that the metal strips and dialysis bag are held in place.
7. Observe the reaction.
8. Attach wire leads to the metal strips and connect these to a flashlight bulb.
9. Prepare five more cells, connect them in series, and use the 9-V battery to play a small radio.

Reaction

1. This battery operates because of a transfer of electrons between magnesium and copper.

$$Mg(s) + Cu^{2+}(aq) \rightarrow Mg^{2+}(aq) + Cu(s) + 1.5\ V$$

2. Electrolysis of water produces bubbles of hydrogen and oxygen gas. Therefore, a hole should be in the stopper.
3. A green coating of copper will begin to appear on the magnesium strip after the cell has operated for a while.

Solutions

1. The copper sulfate solution is 0.5 M: Dissolve 80 g of copper sulfate in 1 liter of water.
2. The sodium sulfate solution is 0.5 M: Dissolve 71 g of sodium sulfate in 1 liter of water.
3. Magnesium ribbon: Clean by dipping quickly into dilute HCl.
4. Copper strips: Clean by dipping quickly into dilute HNO_3.

Teaching Tips

NOTES

1. Biologists usually have dialysis tubing.
2. Wet the dialysis tubing before tying off the end and filling the sack with solution.

3. Connect six of these cells by using wire leads with alligator clips. They will operate a 9-V toy, radio, or calculator for several hours.
4. Mg is oxidized at the anode (+); Cu is reduced at the cathode (−).

QUESTIONS FOR STUDENTS

1. Trace the flow of electrons in this reaction.
2. Is this an oxidation–reduction reaction?
3. If so, what is oxidized and what is reduced?
4. What reactions are evident in the cell?
5. Which metal strip is the anode and which is the cathode?

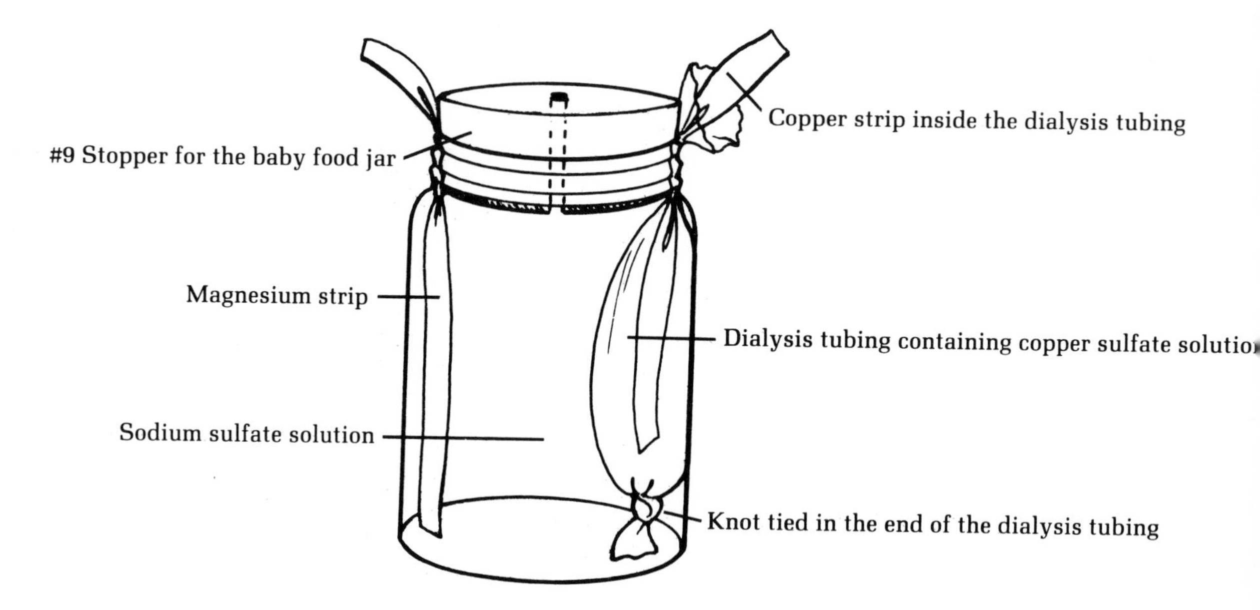

Figure 3. A simple battery made from a baby food jar.

4. Displacement of Tin by Zinc

Mossy zinc is added to a solution in a beaker. After a few seconds, spongy tin is formed and *floats* to the surface of the liquid.

Procedure

1. Place 200 mL of $SnCl_2$ solution in a 600-mL beaker.
2. Add 40 mL of concentrated HCl and stir the solution.
3. Sprinkle about a dozen pieces of granular mossy zinc into the solution; cover the bottom of the beaker with zinc.
4. Spongy tin will immediately form and will soon rise to the surface.

Reactions

This reaction nicely demonstrates two simultaneous reactions:

1. Tin is formed by displacement:

$$Zn(s) + SnCl_2(aq) \rightarrow Sn(s) + ZnCl_2(aq)$$

2. It rises to the surface because hydrogen is also formed:

$$Zn(s) + 2HCl(aq) \rightarrow H_2(g) + ZnCl_2(aq)$$

Solutions

1. The $SnCl_2$ solution is 10%: Dissolve 20 g of $SnCl_2$ in 200 mL of water.
2. The HCl is concentrated.

Teaching Tips

NOTES:

1. The $SnCl_2$ may not be a clear solution, but it will be clear when HCl is added.
2. For a slower reaction, use dilute HCl.
3. Have samples of Zn and Sn available for students to examine and compare to the product.
4. If you want to repeat the demonstration, remove the spongy tin (CAUTION! It may have soaked up concentrated HCl), add a few crystals of $SnCl_2$ and a few milliliters of concentrated HCl.

QUESTIONS FOR STUDENTS

1. Write the chemical equations for these reactions.
2. What is the *electrochemical* reason for the reaction?
3. Why can't the reverse reaction occur spontaneously?
4. Why did the tin float to the surface?

Other Chemical Reactions

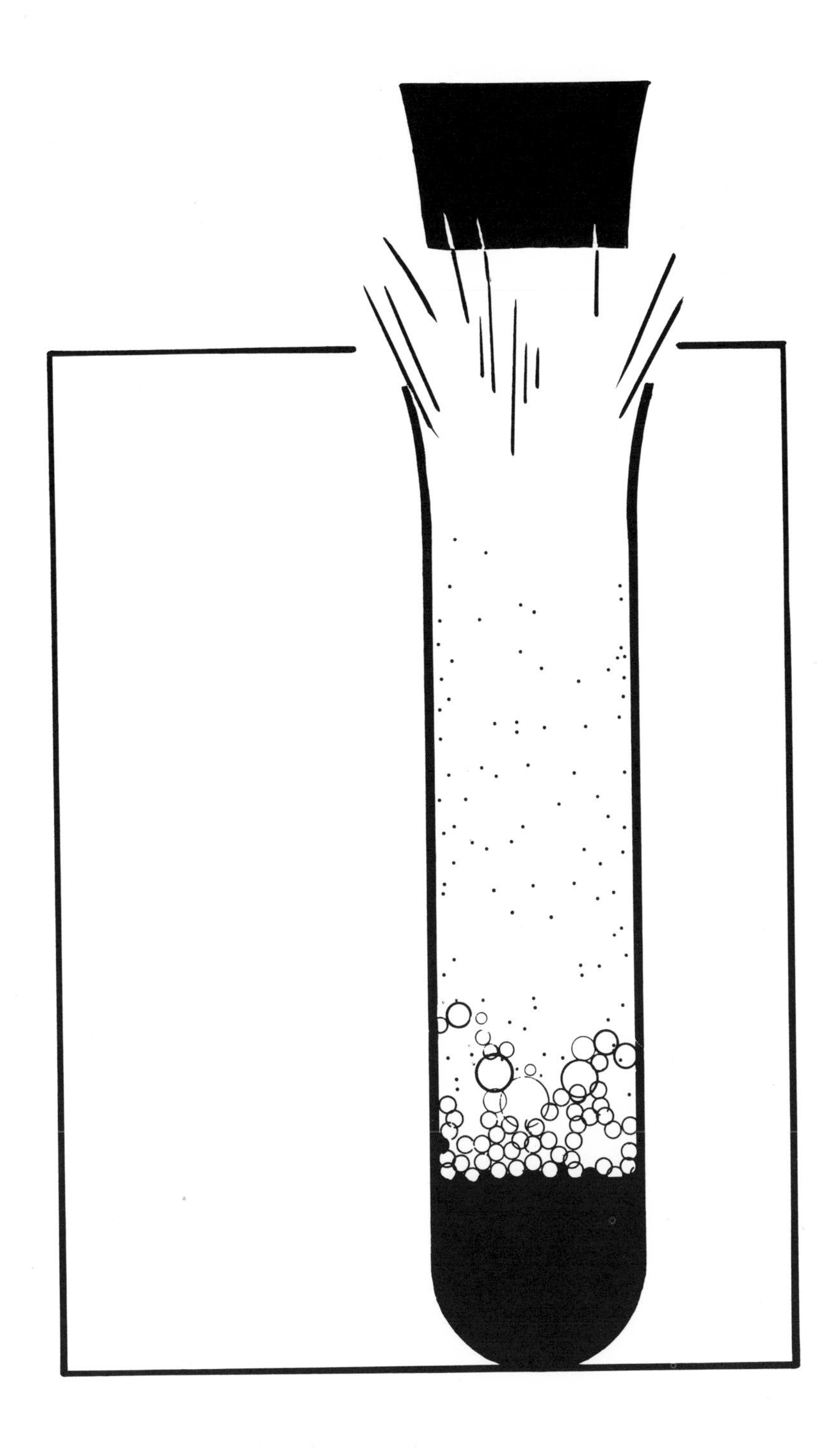

75. Double Displacement: Reaction Between Two Solids

Two white solids are placed in a small beaker and shaken or stirred with a glass rod. A yellow solid is formed.

Procedure

1. Place an equal (approximate) amount of lead nitrate and potassium iodide crystals in a small beaker or petri dish.
2. Students should note that both of these substances are white crystals.
3. Mix the two compounds in the beaker with a glass stirring rod; or, if they are in a petri dish, cover the dish and gently swirl it.
4. Notice the appearance of a yellow product.

Reactions

$$Pb(NO_3)_2(s) + 2KI(s) \rightarrow PbI_2(s) + 2KNO_3(s)$$

(white) (white) (yellow)

Teaching Tips

NOTES

1. This demonstration offers a good opportunity to discuss the importance of surface area in chemical reactions.
2. You might want to pass the containers among the students and let them see the PbI_2 that formed.
3. You can, of course, add equal volumes of KI and $Pb(NO_3)_2$ solutions to obtain the bright yellow PbI_2 precipitate.
4. You might use a large test tube (well stoppered) for mixing the two solids.

QUESTIONS FOR STUDENTS

1. How can a chemical reaction occur between two solids?
2. Write the equation for this reaction.
3. Show how this reaction represents a double displacement.
4. Is a yellow compound the only product of the reaction?

76. Dehydration of Sucrose

Sulfuric acid is poured on sugar in a beaker. In a few minutes, a large carbon *snake* is produced along with a puff of steam and smoke.

Procedure

1. Fill a small beaker about one-third with table sugar (sucrose).
2. Carefully add 5–10 mL of concentrated sulfuric acid. CAUTION! DO NOT STIR.
3. Observe the reactions from a safe distance.

Reactions

1. The acid will dehydrate the sugar and leave only the carbon.
2. Sulfur dioxide and acid droplets are in the fumes from the reaction. THIS DEMONSTRATION SHOULD BE DONE IN A HOOD.
3. The overall reaction is

$$C_{12}H_{22}O_{1}(s) + 11H_2SO_4 \rightarrow 12C(s) + 11H_2SO_4 \cdot H_2O(g)$$

Teaching Tips

NOTES

1. Sucrose is $C_{12}H_{22}O_{11}$.
2. This reaction will char the beaker; therefore, use one that can be discarded after the reaction.
3. Notice that the sugar begins to char prior to the reaction.
4. Handle the carbon snake with TONGS. It will still contain sulfuric acid.

QUESTIONS FOR STUDENTS

1. Write the equation for this reaction.
2. Would this demonstration work with another sugar, say glucose ($C_6H_{12}O_6$)? (Try it!)
3. Why should you *not* handle the carbon mass?

77. Dehydration of *p*-Nitroaniline: Snake and Puff!

A small amount of sulfuric acid is added to a small amount of solid in a beaker. After heating the beaker a few seconds, a giant, sausage-like snake pops out of the beaker with a great deal of smoke. (See Notes.)

Procedure

1. Prepare a paste using a small amount of concentrated sulfuric acid and *p*-nitroaniline.
2. Place a small beaker containing the paste on a wire gauze and heat gently with a burner.
3. When the material begins to char and bubble, reduce the heat and stand back.

Reaction

The sulfuric acid dehydrates the *p*-nitroaniline and produces sulfur dioxide and long, plastic-like polymer snakes.

Teaching Tips

NOTES

1. *p*-Nitroaniline or *p*-nitroacetanilide can be used.
2. This demonstration is spectacular. However, it produces a copious amount of smoke that the class may find irritating. It should be performed outdoors, or at the end of a class when the room can be evacuated.

QUESTIONS FOR STUDENTS

1. What is the appearance of the residue? What does it look like?
2. What is the probable element in the residue?
3. What does dehydration mean?
4. What is the gas that helped form the *snake*?
5. Why do you suppose the snake was produced suddenly, rather than a little at a time?

78. Synthesis of Nylon

Two liquids are mixed in a small beaker and nylon is formed at their interface. The nylon is pulled from the beaker; a continuous thread 10–15 feet long is formed.

Procedure

1. Place 5 mL of solution *A* in a small beaker.
2. Place 5 mL of solution *B* in a second beaker.
3. Slowly add solution *A* to *B* by pouring it down the side of the beaker. DO NOT STIR OR MIX.
4. A film will form at the interface of the two solutions.
5. Carefully hook the film with a bent paper clip and pull the film from the beaker.
6. Continue pulling until the solutions are exhausted.

Reactions

$$x\mathrm{H{-}NH(CH_2)_6NH{-}H} \quad (\text{hexamethylenediamine or 1,6-hexanediamine}) + x\mathrm{Cl{-}\overset{O}{\overset{\|}{C}}(CH_2)_4\overset{O}{\overset{\|}{C}}{-}Cl} \quad (\text{adipoyl chloride}) \xrightarrow{\text{excess amine}} \mathrm{({-}NH(CH_2)_6NH{-}\overset{O}{\overset{\|}{C}}(CH_2)_4\overset{O}{\overset{\|}{C}}{-})_x} \quad (\text{Nylon 66})$$

Solutions

1. Solution *A*: Prepare a 0.5 M basic solution of hexamethylenediamine (or 1,6-diaminohexane) as follows: Warm the solid until it melts. Weigh 5.81 g in a small beaker. Transfer to a large beaker and dilute to 100 mL with 0.5 M NaOH solution (20 g of NaOH per liter).
2. Solution *B*: Adipoyl chloride, 0.25 M: Weight 4.58 g and dilute to 100 mL with hexane.

Teaching Tips

NOTES

1. The nylon can be colored by adding 1 drop of methyl red or bromcresol blue to the solutions.
2. If students want to keep the nylon, it should be washed several times with water.
3. A different nylon, Nylon 610, can be prepared as follows:

   a. Solution *A*: 1 mL of sebacoyl chloride in 50 mL CCl_4.
   b. Solution *B*: 2.5 mL of hexamethylene in 25 mL of water.
   c. Mix two volumes of *A* and one volume of *B*.
   d. The reaction for producing Nylon 610 is

$$H_2N(CH_2)_6NH_2 + ClC(=O)(CH_2)_8C(=O)Cl \xrightarrow{NaOH} H[NH-(CH_2)_6NHC(=O)-(CH_2)_8C(=O)]_xCl$$

Nylon 610

QUESTIONS FOR STUDENTS

1. Why is this called Nylon 66?
2. Why is the nylon synthesized at the interface of the two liquids?
3. What is a polymer?

79. Synthesis of Rayon

The cellulose in filter paper is dissolved and, upon acidification, regenerated as rayon.

Procedure

1. Tear a piece of filter paper into a dozen or so small pieces and place them in a small beaker.
2. Add precipitated copper hydroxide (see Solutions) to the beaker. Add enough concentrated NH_4OH to cover the bottom of the beaker—this reaction forms tetraamminecopper(II) hydroxide.
3. Stir the mixture until the filter paper completely dissolves. (This step may take 30 min.) Add a little more NH_4OH, if necessary.
4. Fill a dropper with the dissolved paper, or use a small syringe.
5. With even pressure, gently squirt the blue solution into a beaker of sulfuric acid. Hold the tip of the dropper or syringe beneath the surface of the acid.
6. Observe the formation of rayon. When this substance turns white, it can be removed, washed, and dried.

Reactions

1. This process for making rayon is the cuprammonium process.
2. Cellulose in filter paper is dissolved by tetraamminecopper(II) hydroxide, $Cu(NH_3)_4(OH)_2$.
3. Upon acidification, cellulose is regenerated into hardened filaments of rayon.

Solutions

1. The precipitated copper hydroxide is made as follows. (Do this precipitation in the HOOD. Strong fumes from ammonium hydroxide may be irritating.)

   a. Fill a 500-mL beaker about one-fourth full with water.
   b. Add copper sulfate and stir until the solution is saturated.
   c. With constant stirring, add concentrated ammonium hydroxide drop by drop. A light blue color will form and a precipitate will appear. (Do not let the solution become dark blue. If it does, start over.) Filter. Wash the precipitate with cold water. Discard the filtrate.

2. The sulfuric acid is 2 M (see Appendix 2).

Teaching Tips

NOTES

1. Chopped-up wood is the commercial source of cellulose.
2. This demonstration projects well—use petri dishes and an overhead projector.
3. Other, better ways are used to make rayon. In the viscose process, cellulose is treated with sodium hydroxide and then carbon disulfide. The cellulose is thus converted to viscose, a viscous yellow fluid. After aging, the viscose is squirted into dilute sulfuric acid and hydrogen sulfate solutions, where if forms rayon. In the cellulose acetate method, cellulose is treated with glacial acetic acid and acetic anhydride. The product is hydrolyzed and then dissolved in acetone. This viscous solution is forced through a spinneret into warm air to form cellulose acetate rayon.

QUESTIONS FOR STUDENTS

1. What is the appearance of the rayon?
2. What dissolves cellulose?
3. Could you use sources of cellulose other than filter paper? Try it!

80. Synthetic Rubber

Two solutions are heated. When cooled, a rubbery mass precipitates. When squeezed and formed into a ball, this material bounces and shows other properties of rubber.

Procedure

Do this reaction in a hood.

1. First, prepare a solution of sodium polysulfide. See Solutions.
2. Place the dark-brown polysulfide in a large beaker.
3. Heat the solution to 65–70 °C and stir in 1.0 g of $Mg(OH)_2$. Continue stirring until the magnesium hydroxide dissolves.
4. Slowly add 25–30 mL of ethylene chloride. You will see an immediate reaction and the evolution of heat. Watch the temperature and do not let it exceed 80 °C! Stir this solution until the color changes from dark brown to a cloudy light brown. Continue to stir for about 15 min.
5. Remove the beaker from the heat and allow it to cool by standing at room temperature.
6. The polysulfide rubber should settle to the bottom of the beaker. Pour off the liquid and rinse the rubber several times with water.
7. The material should coagulate into one large mass. If it doesn't, add ∼ 10 drops of dilute HCl, stir, and rinse again.
8. Remove the rubber, squeeze it to remove water, form it into a ball, and bounce it on the lab bench!

Reactions

1. NaOH and sulfur, S_8, react to form sodium polysulfide:

$$2Na^{+}(aq) + 1/2S_8(s) \rightarrow NaS(=S)-S(=S)Na(s)$$

2. The reaction of sodium polysulfide with ethylene chloride produces a simple condensation polymer and sodium chloride as a by-product:

$$Cl-CH_2-CH_2-Cl + Na-S(=S)-S(=S)-Na \rightarrow \left[CH_2CH_2S(=S)-S(=S)CH_2CH_2SS\right]_n + NaCl$$

polysulfide rubber

Solutions

1. Sodium polysulfide:

   a. Add 10 g of NaOH to 150 mL of water in a beaker and boil until it dissolves (not long!).
   b. Slowly add, with constant stirring, 20 g of sulfur.
   c. Continue stirring until the solution changes from light yellow to dark brown. This change should take about 15 min.
   d. Allow the solution to settle and cool. Pour off the dark-brown sodium polysulfide.

2. Ethylene chloride

Teaching Tips

NOTES

1. This reaction is of historic interest because it was the first synthetic rubber made in the United States.
2. The material produced in this demonstration is the starting material for the vulcanized rubber process.
3. Notice that a polymer consisting of repeating units of ethane and polysulfide is produced.

QUESTIONS FOR STUDENTS

1. Why is this called *polysulfide* rubber?
2. How are the properties of this product similar to those of ordinary rubber? How are they different?
3. What conditions seem to be important in this reaction (e.g., controlling the temperature)?
4. Write the equation for this reaction.

81. A Chemical Sunset

A petri dish is fitted into a cardboard mask on top of an overhead projector. A solution is added and the lights are dimmed. The white light projected through the solution gradually turns yellow, then red, and finally opaque, a change simulating the colors seen during the sunset.

Procedure

1. Cut a hole the size of a petri dish in a piece of cardboard large enough to cover the top of an overhead projector.
2. Place the petri dish in the hole.
3. Add enough $Na_2S_2O_3$ solution to cover the bottom of the dish.
4. Add about 5 mL of concentrated HCl and quickly stir the solution.
5. Observe the color changes.

Reactions

The reaction produces colloidal sulfur that scatters light as it is being formed and produces the different colors. A natural sunset is observed when light is scattered by dust particles in the atmosphere.

1. HCl reacts with sodium thiosulfate to produce thiosulfuric acid.

$$2H^+(aq) + S_2O_3^{2-}(aq) \rightarrow H_2S_2O_3(aq)$$

2. Thiosulfuric acid decomposes immediately, producing sulfurous acid and sulfur in a colloidal suspension.

$$H_2S_2O_3 \rightarrow H_2SO_3 + \text{colloidal sulfur}$$

3. As the amount of colloidal sulfur increases, more light is blocked and the various colors are produced.

Solutions

1. The sodium thiosulfate pentahydrate solution, $Na_2S_2O_3 \cdot 5H_2O$ is 0.03 M: Dissolve 7 g per liter of water.
2. The hydrochloric acid is concentrated.

Teaching Tips

NOTES

1. Sodium thiosulfate pentahydrate is also known as *hypo* and is used in photography.
2. The formation of colloidal sulfur should take about 25–30 s.
3. You can also perform this demonstration by placing the solutions in a glass container in the beam of light from a slide projector.
4. Try decomposing other thiosulfates and polysulfides with acids to produce the sunset effect.

QUESTIONS FOR STUDENTS

1. What is a colloidal suspension?
2. How is this reaction similar to a natural sunset?
3. What causes the different colors to be produced?
4. Write a chemical equation for this reaction.
5. What other compounds might produce a similar reaction? Try them!

82. Production of Sterno: A Gel

Two beakers contain clear solutions. When the contents of one are poured into the other, a solid gel immediately forms.

Procedure:

1. Place 5 mL of solution *A* in a small beaker.
2. Place 30 mL of solution *B* in a second beaker.
3. Challenge a student to see how many times the contents of the two beakers can be mixed—pour *A* into *B*.
4. When the solutions are first mixed, they immediately form a gel. This gel is *Sterno*.

Reactions

The structure of this gel is unclear. The calcium acetate probably forms a network that traps the ethyl alcohol molecules.

Solutions

1. Solution *A*: The calcium acetate is a saturated solution. Dissolve 35 g of calcium acetate $[Ca(CH_3COO)_2]$ in 100 mL of warm water.
2. Solution *B*: 100% ethyl alcohol works best, but other alcohols also work.

Teaching Tips

NOTES

1. If the gel does not form immediately, you need more calcium acetate in solution.
2. A gel is a colloidal system consisting of a liquid (alcohol) dispersed in a solid (calcium acetate).
3. Other gels include jelly, gelatin, and opal.
4. To show that this product really is canned heat, light the beaker. CAUTION—the flame is almost colorless, but the heat is intense. Turn off the lights to better see the pale-blue flame.
5. Try varying the ratio of calcium acetate and alcohol for maximum effect.

QUESTIONS FOR STUDENTS

1. What is a gel?
2. How is a gel, such as canned heat, produced?
3. What are some properties of this gel?

33. Production of a Foam

Two clear solutions are mixed and a chemical foam is produced.

Procedure

1. Place 50 mL of solution *A* in a 250-mL beaker.
2. Place 50 mL of solution *B* in a second beaker.
3. Pour the contents of *A* into *B* and mix quickly.
4. Invert the beaker to show the stability of the foam.

Reactions

1. This foam is produced by the action of carbon dioxide gas on a detergent solution.
2. Aluminum sulfate, $Al_2(SO_4)_3$ provides the acid component for the demonstration:

$$[Al(H_2O)_x]^{3+}(aq) + H_2O(\ell) \rightleftharpoons H_3O^+(aq) + [Al(OH)(H_2O)_{x-1}]^{2+}(aq)$$

3. $NaHCO_3$ produces HCO_3^-.
4. The products of reactions 2 and 3 react to produce CO_2 gas:

$$HCO_3^-(aq) + H_3O^+(aq) \rightarrow 2H_2O(\ell) + CO_2(gas)$$

Solutions

1. Solution *A*: Place 1.0 g of laundry detergent and 7.0 g of $Al_2(SO_4)_3 \cdot 18H_2O$ in a mortar and grind into a powder. Dissolve the powder in 50 mL of water.
2. Solution *B*: Dissolve 5.0 g of $NaHCO_3$ in 50 mL of water.

Teaching Tips

NOTES

1. A *chemical* foam contains CO_2; a *mechanical* foam contains air.
2. A foam is a colloidal system with a gas dispersed in a liquid.
3. Other foams include whipped cream and shaving cream.

QUESTIONS FOR STUDENTS

1. What reactions led to the production of the foam?
2. How is this reaction similar to that involving the production of CO_2 during the baking process?
3. Describe the foam.
4. Name some other examples of foams.

84. Another Foam

Two clear solutions are mixed and a foam is produced.

Procedure

1. Place 100 mL of aluminum sulfate solution in a large beaker or a large graduated cylinder.
2. Add 100 mL of albumin–sodium bicarbonate solution.

Reactions

1. CO_2 is produced by the action of the acidic aluminum sulfate on the sodium bicarbonate.
2. The CO_2 is trapped by the egg white, and a foam is formed.

Solutions

1. Aluminum sulfate solution: Dissolve 25 g of $Al_2(SO_4)_3$ in 100 mL of water.
2. Albumin–Sodium bicarbonate solution: Add 25 g of sodium bicarbonate and 2 g of egg albumin in 150 mL of warm water. Add the albumin slowly. Stir and heat until most of the solids dissolve. Cool, decant the top 100 mL of solution, and discard the remainder.

Teaching Tips

NOTES

1. A foam is a dispersion of a gas in a liquid.
2. Soapsuds can also be used to trap a foam; albumin makes a more stable foam.
3. See Demonstrations 44 and 83 for other methods of producing foams.

QUESTIONS FOR STUDENTS

1. What is a foam?
2. How was a foam produced in this reaction?
3. What is the role of the albumin?
4. Name some other foams.

85. Surface Tension of Water: The Magic Touch

Powered sulfur is sprinkled on the surface of water in a large beaker. The sulfur floats on the surface; however, when it is touched with the finger the sulfur suddenly cascades to the bottom.

Procedure

1. Sprinkle enough sulfur on the surface of water in a large beaker to lightly cover the surface. Do not use pieces large enough to sink.
2. Invite several students to touch the surface of the water. Nothing will happen.
3. Dip your finger in dishwashing detergent or some other wetting agent.
4. Now, when the surface of the water is touched, the sulfur particles will suddenly fall to the bottom.

Reaction

1. The high surface tension of water acts somewhat like an elastic membrane stretched across the water and prevents the sulfur particles from sinking.
2. A wetting agent lowers the surface tension of water and allows the particles of sulfur to drop through the surface.

Teaching Tips

NOTES

1. Experiment to find the type and amount of wetting agent that works best for you.
2. A single drop of diluted dishwashing detergent, or a grain of powdered detergent will produce the cascade effect.

QUESTIONS FOR STUDENTS

1. Why are wetting agents used in dishwashers?
2. Wetting agents are used in washing powders to produce suds. How does this work?
3. How does the surface tension of water compare with that of other liquids.
4. Would this demonstration work in alcohol?

86. Chemiluminescence: The Firefly Reaction

Two clear solutions are mixed in a darkened room. A luminescent blue color is produced that lasts for several seconds.

Procedure

1. Place 100 mL of luminol in a flask.
2. Darken the room.
3. Add 100 mL of bleach solution to the flask.
4. Observe the reaction.

Reaction

In the presence of an oxidizing agent (bleach), luminol is converted to an excited-state product. This product decays to the ground state with the emission of light.

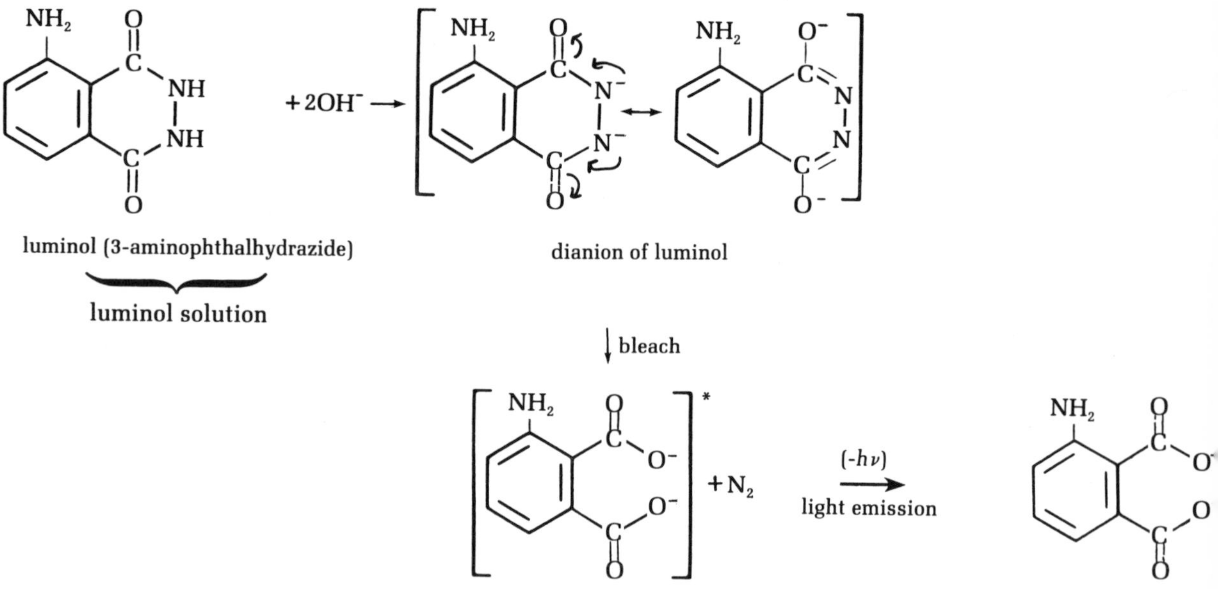

$+ 2OH^- \longrightarrow$

luminol (3-aminophthalhydrazide)

luminol solution

dianion of luminol

bleach

$+ N_2$

$(-h\nu)$ light emission

excited state of 3-aminophthalate ion

ground state of 3-aminophthalate ion

Solutions

1. Luminol solution: Dissolve 0.23 g of luminol in 500 mL of 0.1 M NaOH.
2. Bleach: Dilute laundry bleach (Clorox) 1:10 with water.

Teaching Tips

NOTES

1. Luminol is 5-amino-2,3-dihydro-1,4-phthalazinedione.
2. Bleach is an oxidizing agent. Oxidation occurs without the production of heat.

3. Hydrogen peroxide can also be used as an oxidizing agent. See Demonstration 87.
4. You can make this demonstration more dramatic by pouring the two solutions simultaneously through a funnel into a long coil of glass or tygon tubing.

QUESTIONS FOR STUDENTS

1. Is this an oxidation reaction even though heat wasn't produced?
2. What does the bleach do?
3. How can you explain the production of the blue color?
4. Why is the firefly insect's light yellow?
5. Is biology really chemistry?

87. Chemiluminescence: Two Methods

Method 1

Luminol is added to a jar containing a solution. A brilliant blue color is produced and lasts for several minutes.

Procedure

1. Place 70 g of KOH in a quart jar with a screw cap.
2. Add 60 mL of dimethyl sulfoxide (DMSO).
3. Pass a stream of oxygen gas into the bottle for a few seconds.
4. Cap the jar.
5. When ready for the demonstration add 0.1 g of luminol, shake the jar gently for a few minutes, and pass it around the class.

Reactions

See Demonstration 86 for reactions.

Solutions

1. DMSO is dimethyl sulfoxide. You can buy this chemical at stores where farm supplies are sold. (CAUTION—DMSO penetrates the skin rapidly and carries with it any toxic substances that may be on the surface of the skin.)
2. If you don't have a tank of oxygen, you may be able to borrow one from the shop or maintenance department.

Method 2

Two solutions are mixed; a luminescence is produced for several seconds.

Procedure

1. Mix equal volumes of solution *A* and solution *B*.
2. Observe the chemiluminescence.

Reactions

See Demonstration 86 for reactions.

Solutions

1. Solution *A*: Dissolve 4 g of sodium carbonate in ~ 500 mL of water. Add a small amount of luminol (~ 0.2 g) and stir until it dissolves. Add 0.5 g of ammonium bicarbonate monohydrate. Add 0.5 g of copper sulfate. Add 25 g of sodium bicarbonate. When everything has dissolved, dilute to 1 L.
2. Solution *B*: Dilute 50 mL of 3% H_2O_2 (drugstore variety) to 1 L.

Teaching Tips

NOTES

1. These reactions use oxidizers other than bleach.
2. These reactions produce a more brilliant luminescence that lasts longer than that oxidized by bleach.

QUESTIONS FOR STUDENTS

1. What is the role of hydrogen peroxide?
2. How was the blue color produced?
3. Why was the DMSO solution saturated with oxygen?

88. The Decomposition of Ammonium Dichromate: The *Volcano Reaction*

A small pile of orange crystals is placed in a dish. The chemical is ignited and it burns and produces brilliant sparks and a copious green solid. This display resembles the eruption of a volcano.

Procedure

This reaction should be done in a hood.

1. Place a heaping spoonful of ammonium dichromate in a small pile on a ceramic pad or in an evaporating dish.
2. Ignite the ammonium dichromate by holding a match to the top of the pile or lighting a small strip of magnesium ribbon inserted in the top of the pile.
3. Stand back as the reaction proceeds. It is more dramatic in a darkened room.

Reaction

$$(NH_4)_2Cr_2O_7(s) \rightarrow N_2(g) + 4H_2O(\ell) + Cr_2O_3(s)$$

Teaching Tips

NOTES

1. Do this demonstration in a well-ventilated area.
2. Cover the hood top with newspapers to make cleanup easier.
3. Do not breathe or handle the green residue. It contains chromic oxide, Cr_2O_3. This substance is CARCINOGENIC.
4. Some teachers begin this reaction by adding alcohol and lighting it. This may cause a flash fire, and you should not do it!

QUESTIONS FOR STUDENTS

1. What is produced in this reaction?
2. Is it an oxidation reaction? What has been oxidized?
3. What gas was produced to help expand the residue?
4. Does the match *ignite* the crystals in the same way that a match would *ignite* a piece of paper? Explain.

89. Color Changes in Fe(II) and Fe(III) Solutions

Two sets of four small beakers are arranged on the demonstration table. A solution is poured from a bottle or jug into each beaker and an assortment of colored solutions is produced.

Procedure

FERROUS ION, Fe(II)

1. The beakers are prepared by adding a few crystals (~ 0.5 g) of each of the following solids to ~ 10 mL of water in the beakers: potassium hexacyanoferrate(III) (potassium ferricyanide), $K_3Fe(CN)_6$ (blue); tannic acid, $C_{76}H_{52}O_{46}$ (black); barium chloride, $BaCl_2$ (white); and sodium bisulfite, $NaHSO_3$ (light yellow).
2. The bottle contains iron(II) ammonium sulfate solution, $Fe(NH_4)_2(SO_4)_2 \cdot 6H_2O$.
3. Pour enough solution from the bottle into each beaker to produce the color indicated.

FERRIC ION, Fe(III)

1. The beakers contain a small amount of the following crystals and ~ 10 mL of water: Potassium thiocyanate, KSCN (red); potassium ferrocyanide, $K_4Fe(CN)_6$ (blue); tannic acid, $C_{76}H_{52}O_{46}$ (black); and sodium hydrogen sulfite, $NaHSO_3$ (yellow–orange).
2. The bottle contains iron(III) ammonium sulfate solution, $Fe(NH_4)(SO_4)_2 \cdot 12H_2O$.
3. Pour enough solution from the bottle into each beaker to produce the color indicated.

Reactions

FERROUS ION, Fe(II)

$$3Fe^{2+}(aq) + 2Fe(CN)_6{}^{3-}(aq) \rightarrow \underset{\text{(blue)}}{Fe_3[Fe(CN)_6]_2}$$

$$Fe^{2+}(aq) + \text{tannic acid} \rightarrow \text{Fe(II) tannate (aq)} \xrightarrow{[O]} \underset{\text{(black)}}{\text{Fe(III) tannate(aq)}}$$

$$SO_4{}^{2-}(aq) + Ba^{2+}(aq) \rightarrow \underset{\text{(white)}}{BaSO_4(s)}$$

FERRIC ION, Fe(III)

$$Fe^{3+}(aq) + SCN^-(aq) \rightarrow Fe(SCN)^{2+}$$
(red)

$$Fe^{3+}(aq) + [Fe(CN)_6]^{4-}(aq) \rightarrow Fe^{2+}(aq) + [Fe(CN)_6]^{3-}(aq)$$
(blue)

$$Fe^{3+}(aq) + \text{tannic acid} \rightarrow \text{Fe(III) tannate}(aq)$$
(black)

$$Fe^{3+}(aq) + SO_4{}^{2-}(aq) \rightarrow Fe_2(SO_4)_3$$
(yellow-orange)

Solutions

1. Iron (II) ammonium sulfate, $Fe(NH_4)_2(SO_4)_2 \cdot 6\ H_2O$ and iron(III) ammonium sulfate, $Fe(NH_4)\ (SO_4)_2 \cdot 12\ H_2O$ are dilute solutions. The concentration is not critical—5 to 10 g per liter works well.
2. Try various amounts of solids to get best results.

Teaching Tips

NOTES

1. Tannic acid has the formula $C_{76}H_{52}O_{46}$.
2. The iron(II) ammonium sulfate is readily oxidized to the +3 state.
3. This demonstration is excellent for a *chemical show*. Make up a story to go with the color changes.

QUESTIONS FOR STUDENTS

1. Give the formula for the compounds responsible for each color.
2. Write the equations for each reaction.
3. How could you differentiate between the Fe(II) ion and the Fe(III) ion? Try it! [KSCN gives a red color with Fe(III) but not with Fe(II).]
4. What can you learn about solubility from this demonstration?

90. The Mello-Yello Reaction

Water is poured from a large beaker into five smaller beakers. Solutions that are light yellow, orange, red, black, and finally yellow are produced. When the solution in the final yellow beaker is poured back into the large beaker, the entire contents of the larger beaker turns yellow—hence, *mello-yello*.

Procedure

1. Fill a large beaker, or jar, with water.
2. Pour water into beaker 1. Note the formation of a pale yellow solution. Pour this solution back into the large beaker, and express disappointment that *mello-yello* was not produced.
3. Pour from the large beaker into beaker 2. Note the orange color. Pour this solution back into the large beaker and try again!
4. Pour water into beaker 3, and note the red solution. Pour this solution back into the large beaker.
5. Pour water into beaker 4. A black color is produced. Pour this solution back into the large beaker.
6. Pour water into beaker 5. Mello-yello is produced!
7. When you pour the contents of beaker 5 back into the dark solution in the large beaker, the entire contents of large beaker will turn yellow!

Reactions

Dilute $FeCl_3 \cdot 6H_2O$ is light yellow.

$$Fe^{3+}(aq) + SCN^-(aq) \rightarrow \underset{\text{(red)}}{Fe(SCN)^{2+}}$$

$$Fe^{3+}(aq) + \text{tannic acid} \rightarrow \underset{\text{(black)}}{\text{Fe(III) tannate(aq)}}$$

$$Fe^{3+}(aq) + \text{oxalic acid } (HO_2CCO_2H) \rightarrow \underset{\text{(yellow)}}{Fe_2(C_2O_4)_3 \cdot 5H_2O(aq)}$$

Excess oxalate reacts with all of the Fe^{3+} in the beaker to form yellow iron(III) oxalate.

Solutions

1. Dissolve 30 g of ferric chloride ($FeCl_3 \cdot 6H_2O$) in 100 mL of water. Place 15 drops of this solution in beaker 1.
2. Dissolve 22 g of ammonium thiocyanate (NH_4SCN) in 100 mL of water. Place 2 drops in beaker 2, and place 10 drops in beaker 3.
3. Prepare a saturated solution of tannic acid. Place 12 drops in beaker 4.
4. Prepare a saturated solution of oxalic acid. Place 10 mL (NOT 10 DROPS) in beaker 5.

Teaching Tips

NOTES

1. Place the five solutions in dropper bottles and they can be used over a long period of time.

2. Treat all of these solutions as TOXIC substances.
3. Vary the amounts until you find the combination that gives you the best colors.
4. Tannic acid has the formula $C_{76}H_{52}O_{46}$.

QUESTIONS FOR STUDENTS

1. Write the equations for the reaction that occurs when each small beaker is filled.
2. Are all of these reactions redox reactions?
3. Of those that are redox reactions, what was oxidized and what was reduced?
4. Could the order of *pourings* be changed and the results still be the same? Try it! If so, why? If not, why not?

91. *Magic* Writing Reactions

Three demonstrations are described that involve the sudden appearance of written messages. These reactions are fun and illustrate various chemical reactions resulting in color changes.

Procedures

1. Prepare a writing surface by rubbing a piece of cardboard or posterboard with dry ferric chloride. Paint with colorless solutions of the following chemicals to get the color indicated: Potassium thiocyanate, KSCN (red); potassium ferrocyanide, $K_4Fe(CN)_6$ (blue); and tannic acid, $C_{76}H_{52}O_{46}$ (black).
2. Prepare a writing surface by rubbing poster paper with a mixture of dry potassium ferrocyanide and ferric ammonium sulfate. *Paint* with a brush dipped in water to produce a blue color.
3. Soak a piece of paper in concentrated potassium thiocyanate solution. When the paper is dry, dip a finger into a dilute solution of ferric chloride and draw a bloody picture on the paper.
4. Make up your own variations!

Reactions

$$3SCN^-(aq) + Fe^{3+}(aq) \rightarrow \underset{\text{(red)}}{Fe(SCN)^{2+}}$$

$$Fe^{3+}(aq) + K_4Fe(CN)_6(aq) \rightarrow 4K^+(aq) + Fe^{2+}(aq) + \underset{\text{(blue)}}{[Fe(CN)_6]^{3-}(aq)}$$

$$Fe^{3+}(aq) + \text{tannic acid} \rightarrow \underset{\text{(black)}}{\text{Fe(III) tannate(aq)}}$$

Teaching Tips

NOTES

1. Try other variations.
2. A favorite is the *bloody finger*. Prepare several sheets of paper and store them in a jar. Pretend to cut off the end of a finger. Quickly dip it in the ferric chloride solution, and draw on the paper with *blood*. Wash your hands after this demonstration!

QUESTIONS FOR STUDENTS

1. Write the chemical equations for the reactions.
2. Are these redox reactions?
3. Can you suggest a different *magic* ink?
4. What is the chemistry of your suggestion?

92. Patriotic Colors: Red, White, and Blue

A liquid is poured from a flask into three beakers. Red, white, and blue solutions are produced.

Procedure:

Prepare the flask and beakers as follows:

1. Flask: Fill with 1.0 M ammonium hydroxide solution (see Appendix 2).
2. Beakers:

Beaker 1: 5 drops of alcohol and 5 drops of phenolphthalein

Beaker 2: 5–10 drops of saturated lead nitrate solution (30 g of $Pb(NO_3)_2$ in 100 mL of water).

Beaker 3: 5–10 drops of saturated copper sulfate solution (15 g of $CuSO_4$ in 100 mL of water).

Reactions

1. Ammonium hydroxide reacts with the indicator to give a red color.
2. A double displacement reaction occurs and lead hydroxide precipitates:

$$Pb(NO_3)_2(aq) + 2NH_4OH(aq) \rightarrow \underset{\text{(white)}}{Pb(OH)_2\ (s)} + 2NH_4NO_3(aq)$$

3. A blue complex ion is formed with copper and ammonium ion:

$$Cu^{2+}(aq) + 4NH_4OH(aq) \rightarrow \underset{\text{(deep blue)}}{[Cu(NH_3)_4](OH)_2(aq)}$$

Teaching Tips

NOTES

1. The blue complex ion is tetraamminecopper(II) hydroxide.
2. Adjust the amounts of chemicals to get the desired intensity of color.
3. These colors show up better against a white background.

QUESTIONS FOR STUDENTS

1. Write chemical equations for each of these reactions.
2. Which reaction represents a double displacement?
3. These reactions were produced by adding ammonia. Could they be reversed by adding an acid? Try it!
4. Can you design a demonstration that will produce your school colors?

93. Snakes Alive!

Small, pill-like cones are ignited. A blue flame and long, sausage-like snakes are produced.

Procedure

1. Place the pills on a ceramic pad or on a ring stand base.
2. Light one end.

Reactions

$Hg(NO_3)_2$ and KSCN form the precipitate $Hg(CNS)_2$.

$$Hg(NO_3)_2(s) + 2KSCN(s) \xrightarrow[\Delta]{} 2KNO_3(s) + Hg(CNS)_2(s)$$

Solutions

Prepare the pills as follows

1. Mix a saturated solution of $Hg(NO_3)_2$ and a saturated solution of KSCN (18 g per 10 mL).
2. A white precipitate of mercury thiocyanate will form.
3. After 20–30 min, filter and wash the precipitate.
4. Allow the filtrate to AIR DRY. DO NOT HEAT TO DRY.
5. Make a *dough* of the filtrate by mixing it with a little KNO_3, a little dextrin, a few drops of clear glue, and a few drops of water.
6. Form the dough into small cones, $\sim$ ¼ × ½ in.

Teaching Tips

NOTES

1. Handle $Hg(CNS)_2$ with CARE—it is TOXIC.
2. These are the same *snakes* sold by magic stores—but it is much more fun to make your own!

QUESTIONS FOR STUDENTS

1. Write the equation for this reaction.
2. How can you account for the large mass produced?
3. What type of chemical reaction produces $Hg(CNS)_2$?

94. A Chemical Pop Gun

A solid and liquid are added to a large test tube. The tube is immediately stoppered and pointed away from the class. A loud *POP* results and the stopper is shot across the room.

Procedure

1. Wrap cellophane tape around a large test tube.
2. Place about 10–15 mL of vinegar in the tube.
3. Add a spoonful of sodium carbonate wrapped in a small piece of tissue.
4. IMMEDIATELY stopper the tube and hold it at arm's length. *Point it in a safe direction.*

Reaction

The acetic acid (vinegar) and the carbonate react to form carbon dioxide gas. When confined, the gas exerts enough pressure to force the cork from the tube.

$$Na_2CO_3(s) + 2CH_3COOH(aq) \rightarrow 2NaCH_3COO(aq) + H_2CO_3(aq)$$

$$H_2CO_3(aq) \rightarrow H_2O(\ell) + CO_2(g)$$

Solutions

Use ordinary strength vinegar, which is 5.25% acetic acid.

Teaching Tips

NOTES

1. This simple reaction illustrates various reactions including the action of an acid on a base, the relationship between gas pressure and volume, and a double decomposition.
2. With a little experimenting, you will find just the right combination of vinegar and carbonate for the maximum effect.
3. You can also use a Coke bottle. Although there is little likelihood of the bottle breaking, wrap it with reinforced scotch tape to be on the safe side.
4. If you do not get a loud *POP*, the cork probably isn't fitted tightly enough.

QUESTIONS FOR STUDENTS

1. Why is it necessary for the cork to be firmly fitted in the tube?
2. Write the equation for this reaction.
3. What is the relationship between gas pressure and volume? How does this demonstration show this relationship?

95. Colored Flames

Small piles of chemicals are ignited and various colored flames are produced.

Procedure

1. Place a small pile of the chemicals, combined as indicated (see Table I) for the color desired, in an evaporating dish or on a fireproof surface.
2. Ignite the mixture from a safe distance by using either a burner or a fuse made from filter paper soaked in a KNO_3 solution and allowed to dry.

Table I. Mixtures of Chemicals

Color	*Chemicals*	*Ratio*
Red	Strontium nitrate	5
	Potassium chlorate	4
	Charcoal	1
	Sulfur	1
Purple	Copper sulfate	1
	Potassium chlorate	1
	Sulfur	1
Blue	Copper sulfide	3
	Copper oxide	1
	Mercurous chloride	2
	Potassium chlorate	6
	Charcoal	1
	Sulfur	1
Green	Barium nitrate	10
	Potassium chlorate	3
	Sulfur	2
Yellow	Potassium chlorate	4
	Sodium Oxalate (or chloride)	3
	Charcoal	1
	Sulfur	1
White	Potassium nitrate	6
	Antimony sulfide	1
	Sulfur	1

Note: Ratios are in parts for quick mixing.

Teaching Tips

NOTES

1. If you don't get the intensity of color desired, try varying the amounts of chemicals in the mixture.
2. For best results grind each powder separately and store in a stoppered bottle. DO NOT GRIND POTASSIUM CHLORATE! Mix it lightly with the other compounds.
3. Try creating your own color by recombining chemicals. The following colors can generally be obtained by heating salts of the metals: Scarlet, strontium; red, lithium rubidium; orange-red, calcium; yellow, sodium; yellow-green, barium; green, copper (other than halides); light blue, lead or selenium; deep blue, copper halides blue, cesium; violet, potassium; and white, zinc.

4. Potassium chlorate is a powerful oxidizer. Use it with CAUTION!

QUESTIONS FOR STUDENTS

1. From these recipes, what do you suppose is primarily responsible for the green color? the red color? all of the colors?
2. Does this demonstration have any chemical significance? (Yes, flame tests for some elements depend upon the color of their flame.)
3. Charcoal, sulfur, and potassium chlorate are in most of the mixtures, yet they are not primarily responsible for the color. Why are they there? (Potassium chlorate produces oxygen gas when it decomposes. Sulfur and charcoal form a good ignition mixture to raise the temperature enough to burn the chemical primarily responsible for the color.)

96. Metal Trees

Various metals are placed in flasks containing clear solutions. After a few minutes metallic crystals form on the metal surface, often resembling branches on a tree.

Procedure

You have several ways to demonstrate these reactions:

1. Silver Tree: Place a heavy, coiled copper wire in a 2% silver nitrate solution. A black coating is observed on the wire immediately. After an hour or so, beautiful silver crystals form.
2. Tin Tree: Place a strip of iron, or a coil of iron wire, in a container of tin chloride solution. Observe the formation of tin crystals.
3. Lead Tree: Place a strip of zinc in a 5% solution of lead acetate. Lead crystals form on the zinc strip.

Reactions

$$Ag^{2+}(aq) + Cu(s) \rightarrow Ag(s) + Cu^{2+}(aq)$$

$$Sn^{2+}(g) + Fe(s) \rightarrow Sn(s) + Fe^{2+}(aq)$$

$$Pb^{2+}(aq) + Zn(s) \rightarrow Pb(s) + Zn^{2+}(aq)$$

Solutions

1. The silver nitrate solution is 2%: Dissolve 4 g of $AgNO_3$ in 200 mL of distilled water.
2. The tin chloride solution is 5%: Dissolve 10 g of $SnCl_2$ in 200 mL of water.
3. The lead acetate solution is 5%: Dissolve 10 g of $Pb(CH_3COO)_2$ in 200 mL of water.

Teaching Tips

NOTES

1. Each of these reactions involves the formation of a metal by a displacement reaction.
2. Oxidation–reduction should also be pointed out to the students.
3. Although the silver tree is beautiful and classic, silver is quite expensive. Use sparingly and take necessary precautions when using silver nitrate.
4. Although a reaction is apparent in a few minutes in each of these reactions, complete crystal formation may take several hours.

QUESTIONS FOR STUDENTS

1. Write the equation for each reaction demonstrated.
2. Can you predict what will happen in each reaction?
3. How are these reactions related to the electromotive series of elements?
4. Devise a demonstration to show tree formation from a metal and a solution. Try it!

97a. The Common Ion Effect: First Demonstration

Two Coke bottles are fitted with balloons. Each bottle is filled one-fourth with acetic acid. Solid sodium acetate is added to one bottle. A 3.0-g sample of magnesium metal is added to each bottle, and the balloons are attached. The balloon on the bottle containing the sodium acetate requires longer to inflate than does the balloon on the other bottle.

Reactions

1. Hydrogen is produced in each bottle as a result of the action of acid on magnesium metal.

$$2CH_3COOH(aq) + Mg(s) \rightarrow H_2(g) + Mg(CH_3COO)_2(aq)$$

2. The concentration of hydrogen from the acid has been reduced by the presence of additonal acetate ion.

$$2CH_3COOH(aq) \rightleftharpoons H_2(g) + \underset{\text{(increased)}}{2CH_3COO^-(aq)}$$

3. The rate of the reaction is directly proportional to the concentration of hydronium ion. Therefore, the rate of hydrogen production in the *buffered* solution is less than in the bottle containing only acetic acid.
4. Both balloons eventually reach the same size: thus, the total amount of available hydrogen was not changed in the buffered bottle—only the rate of formation.

Solutions

1. The acetic acid solution is ∼ 2 M. Try using vinegar (5%) diluted one-half (see Appendix 2).
2. Use about 3 g of magnesium powder.
3. Use about 15 g of sodium acetate. (Shake the bottle to dissolve the sodium acetate in acetic acid.)

Teaching Tips

NOTES

1. The balloons MUST be placed on the bottles quickly and simultaneously. Have students help, and go through a few dry runs.
2. Secure the balloons to the bottles with tape or twist ties.
3. It may take 5–8 min for both balloons to reach the same size. Use this time to discuss the common ion effect.

QUESTIONS FOR STUDENTS

1. Write the equation for the reaction in each bottle.
2. What effect does the added acetate have on the rate of hydrogen formation? On the total amount of hydrogen formed?
3. Is this reaction an oxidation–reduction reaction? If so, what is oxidized?

97b. The Common Ion Effect: Second Demonstration

This demonstration is a variation of the reaction in Demonstration 97a, but the gas produced in the reaction causes the production of a foam.

Procedure

1. Obtain two large (500 mL) graduated cylinders.
2. Place a small amount of precipitated calcium carbonate (about a heaping teaspoon works well) in each cylinder.
3. Add 100 mL of acetic acid solution 1 to one cylinder.
4. Add 100 mL of acetic acid solution 2 to the other cylinder.
5. Note the height of foam produced in each cylinder. Mark the level with a wax pencil or with tape.

Reactions

1. Calcium carbonate reacts with acetic acid to produce CO_2 gas.

$$CaCO_3(s) + 2CH_3COOH(aq) \rightarrow Ca^{2+}(aq) + 2(CH_3COO^-)(aq) + H_2O(\ell) + CO_2(g)$$

2. The height of foam produced is assumed to be proportional to the amount of gas produced, and the rate of foam production is proportional to the rate of carbon dioxide production.

Solutions

1. Acetic acid solution 1: 100 mL of 2–3 M acetic acid (see Appendix 2).
2. Acetic acid solution 2: 100 mL of 2–3 M acetic acid to which a heaping spoonful of solid sodium acetate has been added (shake to dissolve).

Teaching Tips

NOTES

1. The cylinder containing acetic acid with sodium acetate will produce hydrogen slower than the other cylinder because of the additional acetate ion. Thus, rapid foam production will not occur.
2. The total amount of foam in each cylinder will be the same, because the equilibrium in the *buffered* solution shifts to ultimately produce the same amount of hydrogen ion as the *unbuffered* solution.
3. Try varying the amount of calcium carbonate and acetic acid to achieve best results.
4. You must use laboratory grade (precipitated) calcium carbonate, not the purified analytical grade variety. This grade will be indicated on the label on the jar.
5. This demonstration offers an excellent opportunity to discuss buffering in acid–base reactions.

QUESTIONS FOR STUDENTS

1. What effect does additional acetate ion have on the production of foam?
2. Write the equation for this reaction.

3. What is the *common ion* effect?
4. When a potassium chloride solution is added to a clear solution of potassium perchlorate, potassium perchlorate precipitates as a milky-white substance. Try it! Explain this result. What is the common ion?

98. The Common Ion Effect: Lead Chromate

A beaker contains a solution of lead chromate. This solution is divided into three smaller beakers. A few drops of lead nitrate are added to the first beaker and a precipitate forms. A few drops of sodium nitrate are added to the second beaker, but no reaction occurs. Some potassium chromate is added to the third beaker and a precipitate forms.

Procedure

1. Fill a large beaker with a solution of $PbCrO_4$.
2. Divide this solution into three smaller beakers.
3. To the first beaker, add $Pb(NO_3)_2$ solution from a dropper. Note the formation of a precipitate.
4. Is the precipitate due to increase NO_3^- concentration? Add $NaNO_3$ to the second beaker to show that no reaction occurs.
5. To the third beaker, add K_2CrO_4 solution, drop by drop. Note the formation of a precipitate.

Reactions

1. In beaker 1, $PbCrO_4$ precipitates from the saturated solution, because Pb^{2+} ion is common to both solutions.

$$PbCrO_4(aq) \rightleftharpoons \underset{\text{(increase)}}{Pb^{2+}(aq)} + CrO_4^{2-}(aq)$$

2. No precipitate occurs in beaker 2. Thus, the precipitate in beaker 1 was NOT due to the added NO_3^- ion.
3. In beaker 3, a precipitate occurs because the CrO_4^{2-} ion is common to both solutions.

$$PbCrO_4 \rightleftharpoons Pb^{2+}(aq) + \underset{\text{(increase)}}{CrO_4^{2-}(aq)}$$

Solutions

1. The $PbCrO_4$ solution is saturated (it is only slightly soluble in water).
2. The $Pb(NO_3)_2$ solution is 1.0 M: Dissolve 3.2 g of lead nitrate in 10 mL of water.
3. The K_2CrO_4 solution is 1.0 M: Dissolve 1.9 g of potassium chromate in 10 mL of water.
4. The $NaNO_3$ solution is about 1.0 M: Dissolve 1.0 g in 10 mL of water.

Teaching Tips

NOTES

1. This demonstration projects well. Use petri dishes and an overhead projector.
2. Use this demonstration to begin a discussion of solubility. LeChatelier's principle, or reactions between ions.
3. The lead chromate solution should be saturated; the concentration of the others is not critical.

QUESTIONS FOR STUDENTS

1. Write the equation for each of the reactions.
2. Define the *common ion* effect.
3. Why is a saturated solution of lead chromate used?
4. What is the *solubility product constant,* K_{sp}? How is it related to the common ion effect?

99. The Common Ion Effect: Ammonium Hydroxide and Ammonium Acetate

A large beaker contains a light red solution of $NH_4^+(aq)$. A small amount of solid NH_4CH_3COO is added and the solution becomes colorless.

Procedure

1. Add 250 mL of water and a few drops of phenolphthalein solution to a beaker.
2. Add NH_4OH, drop by drop, until the color just changes to light red.
3. Add a small amount of solid ammonium acetate or ammonium chloride.
4. Note the change in color from red to colorless.

Reactions

$$NH_3(aq) + H_2O(\ell) \rightleftharpoons NH_4^+(aq) + OH^-(aq)$$

1. Adding ammonium chloride increases the concentration of NH_4^+.

$$NH_4Cl(s) \rightarrow NH_4^+(aq) + Cl^-(aq)$$

2. As additional NH_4^+ is added, the equilibrium shifts to the left, and the concentration of OH^- is reduced to the point that the indicator, phenolphthalein, turns colorless (below pH 9.5).

$$NH_3(aq) + H_2O(\ell) \rightleftharpoons \underset{\text{(increased)}}{NH_4^+(aq)} + OH^-(aq) + Cl^-(aq)$$

Teaching Tips

NOTES

1. You can add sodium acetate, or sodium chloride, to the solution to show that the precipitate is not due to the effect of the acetate, or the chloride.
2. Try other indicators.

QUESTIONS FOR STUDENTS

1. Define the *common ion* effect.
2. What is the common ion in this demonstration?
3. Write the equation for this reaction.
4. What would happen to the equilibrium if NaOH is added? Try it!

Smoke, Fire, and Explosions

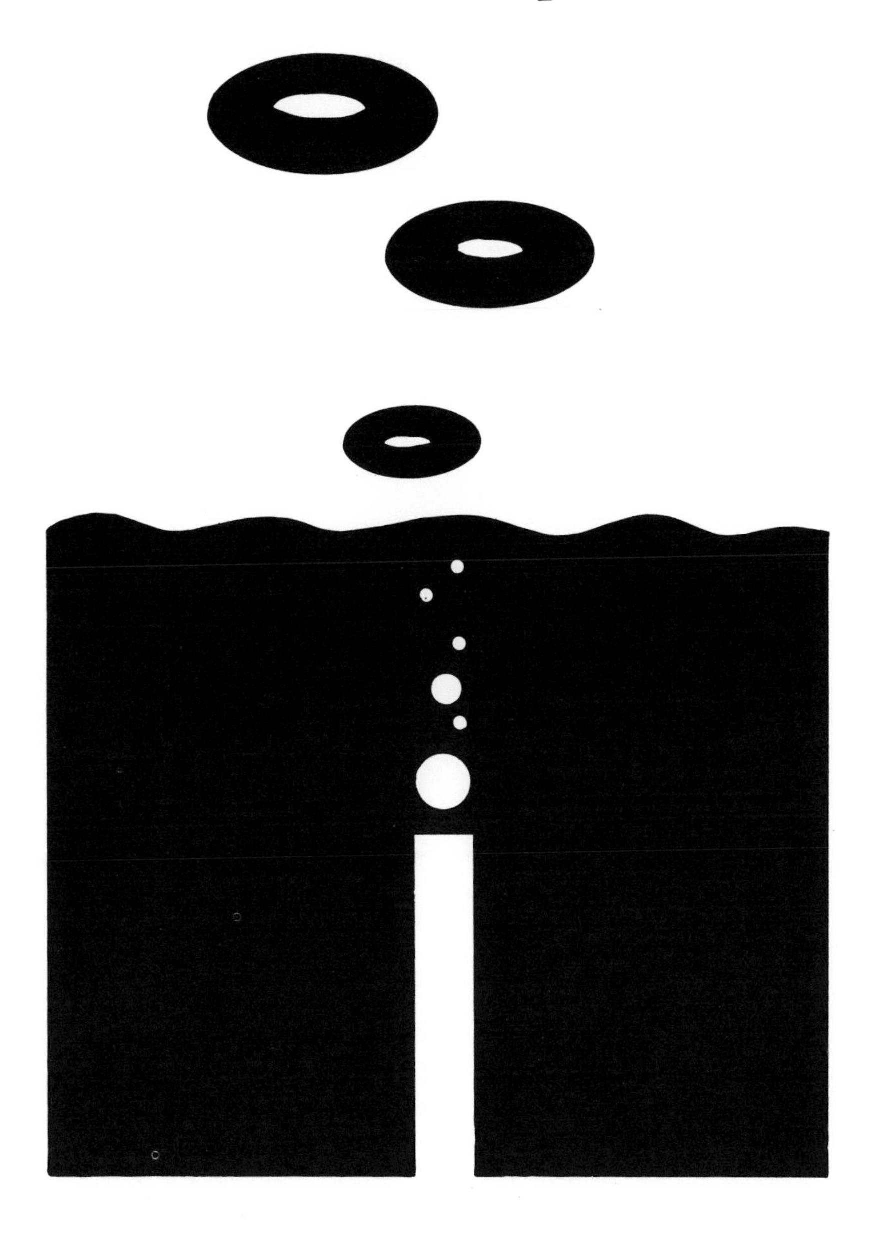

100. Instant Fire

A drop of concentrated sulfuric acid is added to a small amount of a mixture in an evaporating dish. A large and intense flame is instantly produced.

Procedure

1. Mix (DO NOT GRIND) 1 spoonful of granular potassium chlorate ($KClO_3$) with 1 spoonful of table sugar.
2. Place the mixture in a pile in the center of an evaporating dish.
3. CAREFULLY add 1 drop of concentrated sulfuric acid to the top of the pile.
4. STAND BACK!

Reactions

1. This rapid oxidation reaction produces a large amount of heat.
2. The mechanism of this reaction is not well understood. Chloric acid, $HClO_3$, may be produced. This acid is a very active oxidizing agent that readily decomposes the sugar in the presence of sulfuric acid.
3. Sulfuric acid is a powerful dehydrating agent that extracts hydrogen and oxygen as water from the sugar.

Solutions

The sulfuric acid is concentrated.

Teaching Tips

NOTES

1. The potential danger in this demonstration is not in the fire and smoke, but the handling of potassium chlorate. NEVER GRIND POTASSIUM CHLORATE. IT MAY EXPLODE. Improper handling of potassium chlorate is one of the most common causes of laboratory accidents.
2. Clear the area around the evaporating dish, because there may be spattering of sparks.
3. Carefully add the 1 DROP OF ACID, holding the dropper at arm's length. You might want to use a dropper made from a 60-cm length of 0.8-mm o.d. glass tubing.

QUESTIONS FOR STUDENTS

1. Propose a mechanism for this reaction.
2. What is the role of potassium chlorate?
3. What is the role of sugar?
4. What products are formed?
5. What initiates the reaction? How?

101. Production and Spontaneous Combustion of Acetylene

Small lumps of calcium carbide are dropped into a beaker in which chlorine gas is being produced. Bubbles of acetylene form and immediately burst into flame.

Procedure

Do this in a hood.

1. Obtain a 500-mL beaker and a piece of cardboard large enough to cover it.
2. Place 25 mL of sodium hypochlorite (laundry bleach) in the beaker.
3. Add 10 mL of hydrochloric acid.
4. Cover the beaker. Note the reaction and gas production.
5. After a couple of minutes, remove the cover and drop in a few small lumps of calcium carbide. Cover the beaker.
6. Observe the reaction.

Reactions

1. Production of chlorine gas:

$$ClO^-(aq) + Cl^-(aq) + 2H^+(aq) \rightarrow Cl_2(g) + H_2O(\ell)$$

2. Production of acetylene:

$$CaC_2(s) + 2H_2O(\ell) \rightarrow C_2H_2(g) + Ca^{2+}(aq) + 2OH^-(aq)$$

3. The tendency of $Cl_2(g)$ to remove hydrogen is so great that it reacts spontaneously with acetylene, C_2H_2, to produce an exothermic reaction.

$$C_2H_2(g) + Cl_2(g) \rightarrow 2HCl + 2C + \text{heat}$$

Solutions

1. The HCl solution is 6 M (see Appendix 2).
2. Sodium hypochlorite: Laundry bleach is 5.25% NaOCl.

Teaching Tips

NOTES

1. Take precautions against the Cl_2 gas produced in the reaction. Use a hood, or well-ventilated area.
2. Use SMALL pieces of calcium carbide.
3. This reaction shows the oxidizing power of Cl_2 and the reactivity of acetylene.
4. The structure of acetylene, C_2H_2, is $H\text{–}C \equiv C\text{–}H$.
5. This reaction makes a great deal of black smoke!

QUESTIONS FOR STUDENTS

1. What types of bonds between orbitals do you find in acetylene? (One sigma bond and two pi bonds.)
2. When turpentine, $C_{10}H_{16}$, is placed in chlorine gas, it burns immediately. Can you explain this?
3. Give the equations for the reactions in this demonstration.
4. Is this an oxidation–reduction reaction? What is oxidized and what is reduced?

102. Oxidation of Phosphorus: *Barking Dogs*

Several empty glasses of varying sizes are covered with a piece of filter paper. A few drops of solution are placed on each filter paper. In a few seconds, a small explosion occurs in each glass, and various pitches and *woofs* resembling those of barking dogs are produced.

Procedure

1. Select several glass tumblers or beakers of various sizes and shapes. Place a safety shield in front of the containers.
2. Place a filter paper on top of each container.
3. Drop 5–10 drops of phosphorus–carbon disulfide solution in the middle of each paper.
4. The reaction will occur in a few minutes.

Reactions

1. Carbon disulfide evaporates down into the glass container and forms an explosive mixture with air.
2. Phosphorus on the filter paper ignites spontaneously and ignites the explosive mixture in the beaker.

$$P_4(s) + 5O_2(g) \rightarrow P_4O_{10}(s) + \text{heat}$$

$$CS_2(g) + 3O_2(g) \rightarrow CO_2(g) + 2SO_2(g) + \text{heat}$$

Solutions

Prepare the phosphorus–carbon disulfide solution as follows:

1. Dissolve 2 g of yellow or white phosphorus in 10 mL of CS_2, carbon disulfide.
2. CUT YELLOW PHOSPHORUS ONLY UNDER WATER. HANDLE ONLY WITH FORCEPS AND GLOVES.
3. Place the solution in a dropper bottle. Put the dropper bottle inside a can for storage. Handle with care.

Teaching Tips

NOTES

1. Don't add more than the recommended amounts of P–CS_2 to the filter paper!
2. Small bits of sulfur may be deposited on the walls of the containers after the reaction as a result of the incomplete combustion of the gases.
3. A great deal of chemistry is demonstrated by this method:
   a. The solubility of phosphorus in carbon disulfide.
   b. The evaporation of CS_2 that leaves P on the filter paper.
   c. The diffusion of heavier CS_2 into the container.
   d. The spontaneous oxidation of phosphorus.
   e. The explosion of a gaseous mixture.

QUESTIONS FOR STUDENTS

1. Why is the phosphorus dissolved in carbon disulfide?
2. Propose equations for these reactions.
3. Why doesn't the CS_2 evaporate from the filter paper into the air?
4. How do you account for the various *pitches*?

103. An Explosion: The Rapid Oxidation of Phosphorus

A small pile of potassium chlorate that has been treated with a phosphorus–carbon disulfide solution is placed on the metal base of a ringstand. After a few minutes, the pile is touched with the metal rod of the ringstand and a loud explosion results.

Procedure

1. Place a small pile of potassium chlorate, $KClO_3$, NO LARGER THAN AN ASPIRIN TABLET, on the metal base of a ring stand.
2. Add 2-4 drops of phosphorus-carbon disulfide solution to the top of the pile.
3. After 2-3 minutes an explosion should occur spontaneously. If it doesn't, from a safe distance carefully touch the pile with an iron ring stand rod.
4. This reaction produces a very loud noise, much like that of a shotgun blast. Keep students at a safe distance and warn them of the noise.
5. An alternate procedure is to wrap the potassium chlorate in a piece of tissue and place 4-5 drops of the phosphorus solution on the tissue.

Reaction

Potassium chlorate, $KClO_3$, rapidly oxidizes the phosphorus and produces the explosion.

$$2KClO_3(s) \rightarrow 2KCl(s) + 3O_2(g)$$

$$P_4(s) + 5O_2(g) \rightarrow P_4O_{10}(s) + \text{heat}$$

Solutions

Prepare the phosphorus–carbon disulfide solution as follows:

1. Dissolve 2 g of yellow or white phosphorus in 10 mL of CS_2, carbon disulfide.
2. CUT YELLOW PHOSPHORUS ONLY UNDER WATER. HANDLE ONLY WITH FORCEPS AND GLOVES.
3. Place the solution in a dropper bottle. Put the dropper bottle inside a can for storage. Handle with care.

Teaching Tips

NOTES

1. This P–CS_2 solution can also be used with Demonstration 102.
2. DO NOT EXCEED the recommended amount of potassium chlorate.
3. After you add the P–CS_2 solution to the $KClO_3$ allow time for the solvent (CS_2) to evaporate before the reaction occurs.
4. Ignition temperature for white phosphorus is 35–45 °C. This temperature is approximately the temperature of the skin, so never *touch* white phosphorus.

QUESTIONS FOR STUDENTS

1. Write an equation to represent this reaction.
2. This demonstration uses white phosphorus. What is red phosphorus? How does it differ chemically from white phosphorus?
3. White phosphorus, P_4, will burst into flame spontaneously in air and produce $P_4O_{10}(s)$. Propose an equation for this reaction.
4. Why is there a short delay before this reaction produces an explosion?

104. A Puff of Smoke

A small amount of a solid is placed in a test tube that is clamped in an upright position. A drop of liquid is added. In about 15 s, a reaction produces a sudden puff of smoke.

Procedure

1. Clamp a small test tube (the smaller the better) in an upright position.
2. Add enough benzoyl peroxide to cover the bottom of the tube.
3. Add 1 drop of aniline.
4. The reaction occurs in a few seconds.

Reaction

1. Peroxide is a strong oxidizing agent that produces enough heat to convert aniline into gaseous products, including CO_2 and H_2O.
2. A dark residue, oxidation products of aniline, is left in the tube.

Teaching Tips

NOTES

1. Avoid directly breathing the gases produced in this reaction. Some unreacted aniline may be left in the smoke. Use a HOOD or a well-ventilated area.
2. Take care with benzoyl peroxide, as you would with any strong oxidizing agent.
3. Benzoyl peroxide is often found in acne preparations. It oxidizes bacteria on the skin.

QUESTIONS FOR STUDENTS

1. Look up the structure of aniline.
2. What is aniline used for?
3. What does the benzoyl peroxide do?
4. Propose a mechanism for this reaction.

105. The Methanol Cannon

A corked plastic bottle is clamped into position on a ring stand. An electric spark is applied with a Tesla coil to one of two nails placed in the sides of the bottle. A LOUD explosion occurs, and the cork is propelled across the room.

Procedure

1. Prepare the "cannon" by inserting two large nails into the sides of a heavy plastic bottle. Shampoo or juice bottles work well. The points of the nails should be separated by about ¼ in. to provide a gap. (See Figure 4.)
2. Add about 1 mL of methanol to the bottle.
3. Shake the bottle to vaporize and distribute the methanol.
4. Place a tight-fitting cork in the mouth of the bottle.
5. Securely fasten the bottle on a ring stand by clamping the neck of the bottle with a clamp attached to a ring stand. DIRECT THE MOUTH OF THE BOTTLE UPWARDS AND AWAY FROM STUDENTS.
6. Turn on the Tesla coil and apply a spark from the coil to the head of one of the nails in the bottle.
7. A LOUD explosion will result, and the cork will be propelled across the room.

Reaction

The spark ignites the methanol vapor. A rapid exothermic reaction, carbon dioxide, and water are produced.

$$2CH_4OH(g) + 3O_2(g) \rightarrow 2CO_2(g) + 4H_2O(g)$$

Teaching Tips

NOTES

1. You can also use ethanol.
2. Do not use more than 1.0 mL of alcohol. There will probably be enough vapor left for a second reaction, even with this small amount.
3. After a second firing, rinse the bottle with water and dry before using the cannon again. If you allow the bottle to stand without cleaning for 1–2 days, it will fire again.
4. The blue flame of the reaction is clearly visible in a darkened room.

QUESTIONS FOR STUDENTS

1. Why does such a loud explosion result?
2. Write an equation for the reaction that occurs.
3. Why must the bottle be cleaned before it can be used again?

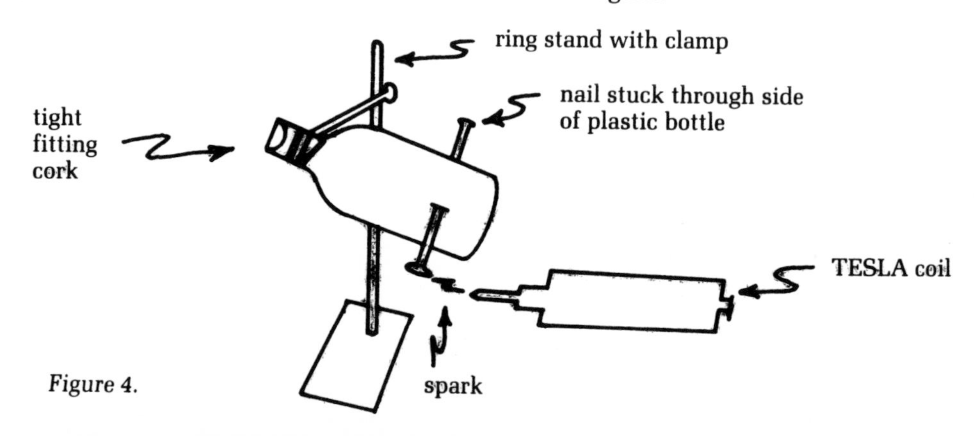

Figure 4.

106. Smoke Rings

White, donut-size smoke rings are produced when phosphine, PH_3, and diphosphorus tetrahydride, P_2H_4, are produced and ignited in the presence of oxygen. DO THIS REACTION IN A HOOD OR WELL-VENTILATED AREA.

Procedure

1. Prepare a reaction vessel by fitting a 500-mL distilling flask with a tight-fitting rubber stopper, or prepare a delivery tube for a 500-mL Florence flask.
2. The tube leading from the flask should bend up into a large container half-filled with water.
3. Attach the flask to a ring stand.
4. Place 100 mL of concentrated NaOH solution in the flask.
5. Add a pea-size piece of white phosphorus.
6. Using a laboratory gas hose, displace the air from the flask and fill with natural gas. Stopper the flask.
7. Heat the flask until the solution begins to boil.
8. Place the delivery tube beneath the water in a large pan or trough.
9. Continue to heat gently.
10. Smoke rings should begin to form.

Reactions

1. Phosphorus and sodium hydroxide produce phosphine:

$$3OH^-(aq) + P_4(s) + 3H_2O(l) \rightarrow 3H_2PO_2^-(aq) + PH_3(g)$$

(hypophosphite)

2. This reaction also produces some diphosphorus tetrahydride (P_2H_4).
3. Because P_2H_4 ignites in air, the air in the flask is displaced with laboratory gas. You could also use nitrogen gas.
4. Phosphine and P_2H_4 bubble through the water in the collecting container. The bubbles ignite on contact with air and form a fog of phosphoric acid, which rises in the form of smoke rings.

$$PH_3(g) + 2O_2(g) \rightarrow H_3PO_4(s) + heat$$

(smoke ring)

Solutions

NaOH is a 40% solution (40 g NaOH in 100 mL.) Be sure to use white or yellow phosphorus.

Teaching Tips

NOTES

1. Observe the usual precautions when using white phosphorus.
2. When you wish to stop the reaction, remove the heat. When the flask cools, slowly add water.

3. PH_3 is similar to NH_3 in structure.
4. Phosphine is a TOXIC substance. In this demonstration, however, all of the phosphine produced reacts to form the phosphoric acid smoke ring. To be on the safe side, use a hood or well-ventilated area.

QUESTIONS FOR STUDENTS

1. Write the equations for the production of phosphine.
2. What causes the smoke rings to form?
3. What role does the laboratory gas serve?
4. How is phosphine similar to ammonia?

107. A Simple Explosive: Nitrogen Triiodide

A small pile of violet-colored crystals is on a table in a roped-off area of the classroom. When this material is touched with a feather attached to a long cane pole, a loud explosion results and violet-colored smoke is produced. (SEE NOTES.)

Procedure

1. Prepare the nitrogen triiodide as follows:

   a. Place 10 g of iodine crystals in a 100-mL beaker.
   b. Add 35–40 mL of concentrated ammonium hydroxide and stir until precipitation ceases.
   c. Filter the solution; catch the nitrogen triiodide on the filter paper. DO NOT LET IT DRY.

2. Prepare the area as follows:

   a. Select a remote area of the laboratory or classroom that can be isolated or roped off.
   b. Carefully scrape the filtrate onto a pad of several filter papers on a table in the roped-off area.
   c. DO NOT ADD MORE MATERIAL THAN WOULD COVER A DIME.
   d. Allow the material to dry, UNDISTURBED, for several hours.

3. When dry, the nitrogen triiodide is extremely sensitive to touch and can be detonated by simply stroking it with a feather attached to a long pole.

Reaction

The iodine crystals react with ammonium hydroxide to form a compound of nitrogen triiodide and ammonia, $NH_3 \cdot NI_3$. This compound is touch-sensitive and undergoes a violent decomposition reaction:

$$8NH_3 \cdot NI_3 \rightarrow 5N_2(g) + 6NH_4I(s) + 9I_2(g)$$

Teaching Tips

NOTES

1. Although there have been reports of the wet material detonating, nitrogen triiodide is quite safe as long as it is HANDLED WHEN IT IS WET. UNDER NO CIRCUMSTANCES SHOULD THE DRY MATERIAL BE HANDLED.
2. This explosive is not powerful, but it is very sensitive. Like any other explosive, it should NEVER BE COMPRESSED, PACKED, OR CONTAINED IN A CLOSED AREA, (e.g., in glassware).
3. Make only as much nitrogen triiodide as you need for a single demonstration. UNDER NO CONDITION SHOULD ANY BE ACCUMULATED OR STORED.
4. The recommended amount (i.e., enough to cover a dime) will produce a very loud, but harmless, explosion. With a little experience, you can vary the amount–use a greater amount only with extreme caution.
5. This demonstration could be saved for an open house or chemical *magic* show. You should NEVER allow students to make this compound!

QUESTIONS FOR STUDENTS

1. Describe the production and explosion of nitrogen triiodide.
2. Write equations for these reactions.
3. Under what conditions would this substance cause damage as an explosive?
4. Propose a model for the structure of NI_3. What can you say about the stability of this structure?

108. An Explosive Combination of Hydrogen and Oxygen

A balloon is taped to a corner of the demonstration desk. The teacher brings a burning match, attached to the end of a meter stick, to the balloon. The balloon explodes.

Procedure

1. The balloon contains a mixture of hydrogen and oxygen gases.
2. Prepare the hydrogen gas as follows (unless you have a lecture bottle of hydrogen): place ∿ 50 mL of 6 M HCl (see Appendix 2) in a Coke bottle; add a few pieces of mossy zinc. As soon as gas production is evident, attach a balloon to the mouth of the bottle.
3. Allow the balloon to inflate as much as possible (about 6 in.).
4. CAREFULLY remove the balloon, without losing any of the hydrogen gas, and inflate it to approximately double its size by adding oxygen from a tank. CAREFUL!
5. Tie the balloon and tape it in a safe place—away from the students.
6. Tape a match to the end of a meter stick. Light the match and touch it—arm's length—to the balloon.

Reaction

$$2H_2(g) + O_2(g) \rightarrow 2H_2O(g) + \text{heat}$$

Teaching Tips

NOTES

1. You may need to experiment to get just the right explosive mixture.
2. This method is one of the safest ways to produce a dramatic explosion in the laboratory.
3. Avoid using concentrated HCl to generate $H_2(g)$.
4. If you have a bell-jar with an opening at the top, fit the top with a one-hole rubber stopper containing the glass tube from a medicine dropper. Cover the bottom of the bell-jar with aluminum foil. Secure the foil to the outside of the jar with masking tape. Punch a small hole in the center of the foil with a pin, and set the bell-jar on bricks or boards to hold the bottom off the demonstration table. Fill the jar with hydrogen through the glass tube at the top. Hold a match to the top of the glass tube. The hydrogen will burn with a pale blue flame. As the hydrogen is consumed, the flame will get smaller and oxygen will be drawn into the jar through the pin-hole in the foil. When the proper combustion mixture of hydrogen and oxygen forms, there will be a loud explosion, and the foil will be blown off the jar.

QUESTIONS FOR STUDENTS

1. Write an equation for this reaction.
2. Hydrogen has a very high heat of combustion when ignited with oxygen. Can you think of any practical uses for this? (Hydrogen will burn in pure oxygen to produce temperatures as high as 2,888 °C! This oxyhydrogen torch will easily cut thick sheets of metal.)
3. Compare the chemical properties of hydrogen with those of oxygen.
4. Is hydrogen now used in blimps and dirigibles? Why? (The *Hindenburg* airship exploded in 1937 as a result of a spark igniting hydrogen gas.)

Appendixes

Appendix 1. The Periodic Table: Electronegativity, Atomic Diameters, and Ionization Energy

KEY: Atomic number (30); Atomic weight (65.37); Atomic diameter Å (2.50); Oxidation states (bold most stable) (2); Ionization potential, ev (9.4); Symbol (Zn); Electronegativity (1.6); Name (Zinc)

Group	Atomic number	Symbol	Name	Atomic weight	Oxidation states	Atomic diameter Å	Ionization potential, ev	Electronegativity
IA	1	H	Hydrogen	1.00797	1	0.64	13.7	2.1
IA	3	Li	Lithium	6.939	1	2.43	5.4	1.0
IA	11	Na	Sodium	22.9898	1	3.08	5.2	.9
IA	19	K	Potassium	39.102	1	4.06	4.4	.8
IA	37	Rb	Rubidium	85.47	1	4.32	4.2	.8
IA	55	Cs	Cesium	132.905	1	4.70	3.9	.7
IA	87	Fr	Francium	(223)	1	–	–	.7
IIA	4	Be	Beryllium	9.0122	2	1.80	9.4	1.5
IIA	12	Mg	Magnesium	24.312	2	2.72	7.7	1.2
IIA	20	Ca	Calcium	40.08	2	3.48	6.2	1.0
IIA	38	Sr	Strontium	87.62	2	3.82	5.8	1.0
IIA	56	Ba	Barium	137.34	2	3.96	5.3	.9
IIA	88	Ra	Radium	(226)	2	–	5.3	.9
IIIB	21	Sc	Scandium	44.956	3	2.88	6.6	1.3
IIIB	39	Y	Yttrium	88.905	3	3.24	6.4	1.2
IIIB	57	La ★	Lanthanum	138.91	3	3.38	5.6	1.1
IIIB	89	Ac ★★	Actinium	(227)	3	–	–	1.1
IVB	22	Ti	Titanium	47.90	4,3	2.64	6.9	1.5
IVB	40	Zr	Zirconium	91.22	4	2.90	6.9	1.4
IVB	72	Hf	Hafnium	178.49	4	2.88	7.0	1.3
IVB	104							
VB	23	V	Vanadium	50.942	5,4,3,2	2.44	6.8	1.6
VB	41	Nb	Niobium	92.906	5,3	2.68	6.9	1.6
VB	73	Ta	Tantalum	180.948	5	2.68	7.9	1.5
VIB	24	Cr	Chromium	51.996	6,3,2	2.36	6.9	1.6
VIB	42	Mo	Molybdenum	95.94	6,5,4,3,2	2.60	7.1	1.8
VIB	74	W	Tungsten	183.85	6,5,4,3,2	2.60	8.0	1.7
VIIB	25	Mn	Manganese	54.938	7,6,4,2,3	2.34	7.5	1.5
VIIB	43	Tc	Technetium	(98)	7	2.54	7.3	1.9
VIIB	75	Re	Rhenium	186.2	7,6,4,2,−1	2.56	7.9	1.9
VIII	26	Fe	Iron	55.847	2,3	2.34	7.9	1.8
VIII	44	Ru	Ruthenium	101.07	2,3,4,6,8	2.50	7.4	2.2
VIII	76	Os	Osmium	190.2	2,3,4,6,8	2.52	8.7	2.2
VIII	27	Co	Cobalt	58.933	2,3	2.32	7.9	1.8
VIII	45	Rh	Rhodium	102.905	2,3,4	2.50	7.5	2.2
VIII	77	Ir	Iridium	192.2	2,3,4,6	2.54	8.7	2.2
VIII	28	Ni	Nickel	58.71	2,3	2.30	7.7	1.8
VIII	46	Pd	Palladium	106.4	2,4	2.56	8.4	2.2
VIII	78	Pt	Platinum	195.09	2,4	2.60	9.2	2.2
IB	29	Cu	Copper	63.54	2,1	2.34	7.8	1.9
IB	47	Ag	Silver	107.870	1	2.68	7.6	1.9
IB	79	Au	Gold	196.967	3,1	2.68	9.3	2.4
IIB	30	Zn	Zinc	65.37	2	2.50	9.4	1.6
IIB	48	Cd	Cadmium	112.40	2	2.96	9.1	1.7
IIB	80	Hg	Mercury	200.59	2,1	2.98	10.5	1.9
IIIA	5	B	Boron	10.811	3	1.64	8.4	2.0
IIIA	13	Al	Aluminum	26.9815	3	2.36	6.1	1.5
IIIA	31	Ga	Gallium	69.72	3	2.52	6.0	1.8
IIIA	49	In	Indium	114.82	3	2.88	5.6	1.7
IIIA	81	Tl	Thallium	204.37	3,1	2.96	6.2	1.8
IVA	6	C	Carbon	12.01115	±4,2	1.54	11.3	2.5
IVA	14	Si	Silicon	28.086	4	2.22	8.2	1.8
IVA	32	Ge	Germanium	72.59	4	2.44	8.0	1.8
IVA	50	Sn	Tin	118.69	4,2	2.82	7.4	1.8
IVA	82	Pb	Lead	207.19	4,2	2.94	7.5	1.8
VA	7	N	Nitrogen	14.0067	±3,5,4,2	1.50	14.6	3.0
VA	15	P	Phosphorus	30.9738	±3,5,4	2.12	11.1	2.1
VA	33	As	Arsenic	74.922	±3,5	2.40	9.9	2.0
VA	51	Sb	Antimony	121.75	±3,5	2.80	8.7	1.9
VA	83	Bi	Bismuth	208.980	3,5	2.92	7.4	1.9
VIA	8	O	Oxygen	15.9994	−2	1.46	13.7	3.5
VIA	16	S	Sulfur	32.064	±2,4,6	2.04	10.4	2.5
VIA	34	Se	Selenium	78.96	−2,4,6	2.32	9.8	2.4
VIA	52	Te	Tellurium	127.60	−2,4,6	2.72	9.1	2.1
VIA	84	Po	Polonium	(210)	2,4	2.92	–	2.0
VIIA	9	F	Fluorine	18.9984	−1	1.44	17.5	4.0
VIIA	17	Cl	Chlorine	35.453	±1,3,5,7	1.98	13.1	3.0
VIIA	35	Br	Bromine	79.909	±1,5	2.28	11.9	2.8
VIIA	53	I	Iodine	126.904	±1,5,7	2.66	10.5	2.5
VIIA	85	At	Astatine	(210)	±1,3,5,7	2.90	–	2.2
VIIIA	2	He	Helium	4.0026		1.86	24.6	
VIIIA	10	Ne	Neon	20.183		1.42	21.6	
VIIIA	18	Ar	Argon	39.948		1.96	15.8	
VIIIA	36	Kr	Krypton	83.80		2.24	14.1	
VIIIA	54	Xe	Xenon	131.30		2.62	12.2	
VIIIA	86	Rn	Radon	(222)		–	10.8	
★	58	Ce	Cerium	140.12	3,4	3.30	–	1.1
★	59	Pr	Praseodymium	140.907	3,4	3.30	–	1.1
★	60	Nd	Neodymium	144.24	3	3.28	–	1.1
★	61	Pm	Promethium	(147)	3	3.26	–	1.1
★	62	Sm	Samarium	150.35	3,2	3.24	–	1.1
★	63	Eu	Europium	151.96	3,2	3.70	–	1.1
★	64	Gd	Gadolinium	157.25	3	3.22	–	1.1
★	65	Tb	Terbium	158.924	3,4	3.18	–	1.2
★	66	Dy	Dysprosium	162.50	3	3.18	–	1.2
★	67	Ho	Holmium	164.930	3	3.16	–	1.2
★	68	Er	Erbium	167.26	3	3.14	–	1.2
★	69	Tm	Thulium	168.934	3,2	3.12	–	1.2
★	70	Yb	Ytterbium	173.04	3,2	–	–	1.2
★	71	Lu	Lutetium	174.97	3	3.12	–	1.2
★★	90	Th	Thorium	232.038	4	3.30		
★★	91	Pa	Profactinium	(231)	5,4	–		
★★	92	U	Uranium	238.03	6,5,4,3	2.84		
★★	93	Np	Neptunium	(237)	6,5,4,3			
★★	94	Pu	Plutonium	(242)	6,5,4,3			
★★	95	Am	Americium	(243)	6,5,4,3	–	–	
★★	96	Cm	Curium	(247)	3	–	–	
★★	97	Bk	Berkelium	(247)	4,3	–	–	
★★	98	Cf	Californium	(249)	3	–	–	
★★	99	Es	Einsteinium	(254)	–	–	–	
★★	100	Fm	Fermium	(253)	–	–	–	
★★	101	Md	Mendelevium	(256)	–	–	–	
★★	102	No	Nobelium	(254)	–	–	–	
★★	103	Lr	Lawrencium	(257)	–			

Appendix 2. Properties and Preparation of Laboratory Acids and Bases

Parameter	*Ammonium Hydroxide (NH_4OH)*	*Acetic Acid ($HC_2H_3O_2$)*	*Hydrochloric Acid (HCl)*	*Nitric Acid (HNO_3)*	*Sulfuric Acid (H_2SO_4)*
Dilute this volume (in milliliters) of concentrated reagent to 1 L to make a 1.0 M solution	67.5	57.5	83.0	64.0	56.0
Dilute this volume (in milliliters) of concentrated reagent to 1 L to make a 3.0 M solution	200	172	249	183	168
Dilute this volume (in milliliters) of concentrated reagent to 1 L to make a 6 M solution	405	345	496	382	336
Normality of concentrated reagent	14.8	17.4	12.1	15.7	36.0
Molecular weight	35.05	60.05	36.46	63.02	98.08
Specific gravity of concentrated reagent	0.90	1.05	1.19	1.42	1.84
Approximate percentage in concentrated reagent	57.6	99.5	37.0	69.5	96.0

Note: To make *normal* solutions, use the same amount of reagent shown. However, to make a normal solution of sulfuric acid, use half the amount of reagent indicated. Example: dilute 28.0 mL of concentrated sulfuric acid to make 1 L of 1.0 N sulfuric acid solution.

Appendix 3. Equipment and Reagent List

This list includes the reagents and the unusual equipment that are needed for each demonstration. Graduated cylinders for measuring amounts and balances for weighing have not been listed. This list is for your convenience. You should always look at the individual demonstration and determine the concentration that works best for you and the purity and age of your stock reagents. Following the equipment names, the numbers in parentheses are the number of pieces needed. Following the reagent names, the letters in parentheses indicate solid (s) or crystal (c).

The numbers in italics indicate the number of the demonstration.

Equipment

Aluminum foil *17*
Anvil and hammer *105*
Ashes and carbon *42*
Aspirator *39*

Baby food jar and stopper *73*
Balloons *3, 4, 9, 97, 108*
Battery jar, 2.0 L *18*
Beaker (2), 150 mL *13, 41*
Beaker (2), 250 mL *12, 43, 47, 49, 51*
Beaker (3), 100 mL *32*
Beaker (3), 250 mL *48*
Beaker (4), 100 mL *57,63*
Beaker (5), 50 mL *90*
Beaker (8), 50 mL *89*
Beaker, 50 mL *27*
Beaker, 150 mL *17, 34*
Beaker, 250 mL *14, 26, 28, 30, 32*
Beaker, 400 mL *36, 40*
Beaker, 500 mL *5, 19, 23*
Beaker, 600 mL *37, 52, 74*
Blocks of plywood *1*
Bottle, 250 mL *50*
Buret *70*

Cellulose, filter paper *79*
Ceramic pad *88, 93, 95, 103*
Coke bottles (2) *97*
Cotton swab *8*
Cups or styrofoam *3*

Deflagrating spoon *21*
Dialysis tubing *73*
Dishwashing detergent *44, 97*
Dropper bottles (5) *90*
Droppers *5, 9, 14, 15, 23, 34, 60*

Elmer's glue *93*
Erlenmeyer flask, 150 mL *16, 33*
Erlenmeyer flask, 250 mL *4, 33, 55*
Erlenmeyer flask, 500 mL *25*
Evaporating dish *56, 60, 88*

Filter paper *102, 107*
Filtering flask, 1.0 L *6, 7*
Florence flask, 500 mL *5, 106*

Glass tube, 3 cm × 50 cm *8*
Glass tubing *5*
Glasses (6) *102*
Graduated cylinder, 50 mL *46*
Graduated cylinder, 500 mL *10, 21, 44, 84*
Graduated cylinder (8), 50 mL *24*
Graduated cylinder (2), 100 mL *8*
Graduated cylinders (2), 500 mL *97*

Hot plate *16, 17, 19, 32, 33, 37, 38, 45, 80*

Icebath *32, 33, 36, 37*
Iron nail *61*

Laundry detergent *83, 84*
Leads and alligator clips *73*
Litmus papers *8*

Magnetic stirrer *52, 53*
Matches *9, 42, 93*
Meter stick and feather *107*
Meter stick *9*
Microspatula *105*

Overhead projector *31, 40*

Paper napkin *62*
Petri dish (3) *34*
Petri dish *31, 40, 54, 61, 65, 67, 81*
Pipet, 25 mL *68*
Poster board *91*

Ringstand and ring *5*
Ringstand and clamp *68*
Round-bottom flask, 500 mL *58, 59*
Round toothpicks *1*
Rubber tubing *6, 11, 36, 39*

Salt brine *38*
Saran Wrap *8*

Side-arm flask, 250 mL 36, 39
Steel wool 6
Styrofoam cup 26, 29
Syringe, 100 mL 11
Syringe 79

Test tube (2) 36
Test tube, 25mm × 200 mm 37
Test tube, 25mm × 250mm 15
Test tube and cork stopper 20
Test tubes (4), 10mm × 75mm 20
Test tubes, plastic 36
Thermometer 15, 16, 26, 29, 30, 80
Tongs 76
Toothpicks 42
Triple-beam balance 10
Trough 6, 10

Wooden block 27, 28
Wooden splint 6

Reagents

Acetic acid, 3 M 97
Acetic acid, glacial 51, 79
Acetic anhydride 79
Acetone 79
Adipyl chloride, 0.25 M 78
Albumin 84
Alka Seltzer tablets 43
Aluminum sulfate, 0.21 M 83
Aluminum sulfate, 0.375 M 84
Ammonium bicarbonate (c) 87
Ammonium chloride (c) 5, 27, 40, 60, 99
Ammonium dichromate (c) 88
Ammonium hydroxide, 6.0 M 24
Ammonium hydroxide, concentrated 8, 37, 55, 79, 99, 107
Ammonium hydroxide, dilute 92
Ammonium Iron(III) sulfate, 0.0002 M 64
Ammonium nitrate (c) 26, 27, 60
Ammonium nitrate, 1.5 M 58
Ammonium thiocyanate (c) 27
Ammonium thiocyanate, 2.9 M 90
Ammonium vanadate (c) 69, 70
Aniline 104
Antimony potassium tartrate, 0.44 M 12
Antimony(III) chloride (s) 40
Antimony(III) sulfide (c) 95

Barium chloride (c) 89
Barium chloride, 1.2 M 12
Barium hydroxide (s) 27
Barium nitrate (c) 95
Barium nitrate, 0.1 M 35
Benzene 21
Benzoyl peroxide 104
Bismuth(III) chloride (s) 40

Bromcresol blue 78
Bromophenol blue 46
Bromthymol blue 24, 25
Bubble solution 2
Butane gas 2
tert-Butyl bromide 22
tert-Butyl chloride 22

Calcium acetate, saturated 16, 82
Calcium carbide (c) 9, 101
Calcium carbonate (s) 97
Calcium chloride (c) 29
Calcium hydroxide, saturated 14
Carbon disulfide 79
Carbon tetrachloride 21
Cerium(IV) sulfate, 0.10 M 70
Charcoal powder 95
Cobalt chloride, 0.5 M 14
Cobalt(II) chloride (c) 6, 18, 19, 45
Cobalt(II) chloride, 0.2 M 34
Cobalt(II) chloride, 0.4 M 32
Cobalt(II) chloride, 2.0 M 20
Cobalt(II) chloride, saturated 38
Cobalt(II) nitrate 0.2 M 6
Cobalt(II) nitrate (c) 18
Cobalt(II) sulfate (c) 28
Copper metal 36, 65, 73
Copper pennies 66
Copper wire, 24 gauge 55, 96
Copper(II) chloride (c) 18
Copper(II) oxide (s) 95
Copper(II) sulfate (c) 79, 87, 95
Copper(II) sulfate, 0.5 M 73
Copper(II) sulfate, concentrated 92
Copper(II) sulfide(s) 95

Dextrin (s) 93
Dextrose (c) 50
Dimethylsulfoxide (s) 87
Dry ice 23, 24

Ethyl alcohol, 100% 82
Ethyl alcohol, 95% 12, 19, 20, 21, 25, 58, 68, 85, 92
Ethylene chloride 80

Ferric chloride, 1.1 M 90
Ferroin, 0.25 M 54
Fructose (c) 58

Glucose (c) 58, 59
Glycerin 56

Hexamethylenediamine (s) 78
Hydrochloric acid, 1.0 M 7, 35
Hydrochloric acid, 3.0 M 80
Hydrochloric acid, 6.0 M 23, 39, 65, 101

Hydrochloric acid, concentrated 8, 13, 32, 34, 40, 74, 81
Hydrogen gas 2, 108
Hydrogen peroxide, 3% 57, 71
Hydrogen peroxide, 6% 45
Hydrogen peroxide, 30% 44, 53

Indigo carmine, 1.0% 50
Iodine (c) 21, 107
Iron wire, 24 gauge 65, 96
Iron(II) ammonium sulfate (c) 89
Iron(III) ammonium citrate (c) 72
Iron(III) ammonium sulfate (c) 57, 89, 91
Iron(III) chloride (c) 18, 91
Iron(III) chloride, concentrated 91
Iron(III) nitrate, 0.2 M 31
Isopropyl alcohol 22, 85

Laundry bleach, 5.25% 6, 7, 30, 86, 101
Lead(II) acetate, 0.15 M 96
Lead(II) chromate, saturated 98
Lead(II) nitrate (c) 75
Lead(II) nitrate, 1.0 M 41, 98
Lead(II) nitrate, concentrated 92
Luminol (s) 86, 87

Magnesium hydroxide (s) 80
Magnesium powder 97
Magnesium ribbon 65, 73
Malonic acid (c) 52–54
Manganese(II) nitrate (c) 18
Manganese(II) sulfate (c) 52, 53
Manganese(II) sulfate, 4.6 M 15
Manganese(IV) dioxide (s) 44
Mercury(I) chloride (s) 95
Mercury(II) chloride (c) 69
Mercury(II) chloride, 0.01 M 48
Mercury(II) chloride, 0.025 M 49
Mercury(II) chloride, 0.1 M 13
Mercury(II) nitrate, saturated 93
Methane gas 2
Methyl alcohol 21, 55
Methyl chloride 21
Methyl red 78
Methylene blue (s) 59

Nickle(II) nitrate (c) 18
Nitric acid, 1.0 M 35
Nitric acid, concentrated 36, 69
p-Nitroacetaniline 77
p-Nitroaniline (s) 77

Oxalic acid, saturated 90
Oxygen gas 87, 107

Petroleum ether 21
Phenol red 24, 25
Phenolphthalein 5, 24, 37, 39, 67, 99
Phosphorus (s), red 105
Phosphorus (s), white or yellow 106
Phosphorus–carbon disulfide 102, 103
Platinum wire 55
Potassium bromate (c) 52
Potassium bromide, saturated 33
Potassium chlorate (c) 46, 100, 103, 105
Potassium chromate, 1.0 M 35, 97, 98
Potassium dichromate, 0.050 M 57
Potassium dichromate, 0.1 M 35, 71
Potassium ferricyanide, 0.7 M 72
Potassium ferrocyanide (c) 89, 91
Potassium ferrocyanide, 0.0002 M 64
Potassium hydrogen sulfate, 0.10 M 64
Potassium hydroxide, 0.50 M 59
Potassium hydroxide, 1.0 M 35
Potassium iodate (c) 53
Potassium iodate, 0.020 M 47
Potassium iodate, 0.070 M 48
Potassium iodide (c) 44, 75
Potassium iodide, 0.100 M 49
Potassium nitrate (c) 93, 95
Potassium perchlorate, saturated 97
Potassium permanganate (c) 56, 68
Potassium permanganate, 0.01 M 62, 63
Potassium permanganate, 0.050 M 57
Potassium thiocyanate (c) 89, 91, 93
Potassium thiocyanate, 0.002 M 31
Potassium thiocyanate, 1.0 M 57

Salt brine, saturated 15
Sebacoyl chloride 78
Silver nitrate, 0.05 M 58
Silver nitrate, 0.12 M 96
Silver nitrate, 0.1 M 34
Soap solution 2
Sodium acetate (c) 97, 99
Sodium acetate, saturated 17
Sodium arsenite, 0.16 M 51
Sodium bicarbonate (s) 84, 87
Sodium bicarbonate, 0.20 M 83
Sodium bisulfide (c) 89
Sodium bromate (c) 54
Sodium bromide, 1.0 M 41
Sodium carbonate (c) 87, 94
Sodium chloride (c) 95
Sodium hydrogen carbonate, saturated 39
Sodium hydrogen sulfite, 0.01 M 63
Sodium hydrogen sulfite, 0.144 M 48
Sodium hydroxide (c) 5, 50, 60, 66, 70, 80
Sodium hydroxide, 0.1 M 86
Sodium hydroxide, 0.5 M 78
Sodium hydroxide, 1.0 M 14, 22, 23, 35
Sodium hydroxide, 2.0 M 25, 63
Sodium hydroxide, 2.5 M 58
Sodium hydroxide, 10 M 106

Sodium metabisulfite (c) *47*
Sodium monohydrogen phosphate (c) *31*
Sodium nitrate, 1.0 M *98*
Sodium oxalate (c) *95*
Sodium potassium tartrate, 0.2 M *45*
Sodium silicate, saturated *18*
Sodium sulfate (c) *33*
Sodium sulfate, 0.5 M *73*
Sodium sulfite (c) *46*
Sodium sulfite, 0.5 M *30*
Sodium thiosulfate, 0.03 M *81*
Sodium thiosulfate, 1.25 M *51*
Sodium (s) *67*
Starch solution, 10% *47*
Strontium nitrate (c) *95*
Sucrose (c) *76, 100*
Sugar cubes *42*
Sulfur powder *85, 95*
Sulfur (s) *80*
Sulfuric acid, 1.0 M *47, 61*
Sulfuric acid, 2.0 M *63, 79*
Sulfuric acid, 4.0 M *46*
Sulfuric acid, 6.0 M *57*
Sulfuric acid, 9.0 M *69, 70*
Sulfuric acid, concentrated *15, 52–54, 61, 76, 77, 100*

Tannic acid (c) *89*
Tannic acid, concentrated *91*
Tannic acid, saturated *90*
Tartaric acid (c) *58*
Thionyl chloride *28*
Thymolphthalein *24*
Tin metal *65, 74*
Tin(II) chloride, 0.10 M *13, 57*
Tin(II) chloride, 0.2 M *96*
Tin(II) chloride, 0.53 M *74*
Turpentine *7, 101*

Universal indicator *22, 23*
Uranyl nitrate (c) *18*

Vinegar *94, 97*

p-Xylene *21*

Zinc dust *60*
Zinc metal *65*
Zinc sulfate *18*
Zinc, granular *66, 69*
Zinc, mossy *74*

Appendix 4. Safety and Disposal

The following information is meant to be a general guide for you to consider before performing a demonstration and disposing of the compounds formed in each demonstration. Obviously, you should not perform a demonstration that uses or produces substances that are on the *banned* list for your school. Check with your district science supervisor or state department of education. You can follow these general rules for disposal of small amounts of waste.

1. Neutralize acids with ammonium hydroxide or sodium hydroxide and flush down the drain.
2. Neutralize bases with acetic acid or dilute hydrochloric acid and flush down the drain.
3. Dissolve inorganic salts in water, neutralize with sodium bicarbonate, and flush down the drain.
4. Allow volatile organic solvents to evaporate in a hood, or burn them in an open fire in a safe area.
5. Place solids in the waste containers, or burn them in an open fire.

If you accumulate chemical waste in large quantity, you should seek professional help with disposal. Contact your local college or university chemistry department, or local industry for advice and assistance.

Although the quantity of chemicals and solutions used in these demonstrations is relatively small, proper disposal procedures should be followed. You may also wish to use the procedures described in this section to discard old and unwanted chemicals from your supply area.

The following disposal procedures follow the recommendations of the Chemical Manufacturers Association, as published in their "Laboratory Waste Disposal Manual." Other excellent sources are: *Hazardous Chemicals, Information and Disposal Guide,* Margaret-Ann Armour, *et al.*, Dept. of Chemistry, Univ. of Alberta, Edmonton, Alberta, T6G 2G2; and Jay Young, "Academic Laboratory Waste Disposal: Yes, You Can Get Rid of That Stuff Legally!", *Journal of Chemical Education*, vol. 60, no. 6, June, 1983. For your convenience, chemicals are divided into inorganic and organic and are listed by major chemical groups.

Inorganic Chemicals

Chemical	Disposal Procedure
Alkali–alkaline earth metals	Mix with DRY sodium carbonate. Place in a dry container and BURN in an open area.
Carbide	Slowly add to water in a hood or open area. When the production of gas ceases, flush the residue down the drain.
Cyanides	Add slowly to a solution of calcium hydroxide and sodium carbonate. Flush the neutralized solution down the drain.
Hydroxides	Dilute in a large container of water. Neutralize with 6 M HCl. Flush down drain.
Inorganic acids	Add slowly to a solution of calcium hydroxide and sodium carbonate. Flush the neutralized solution down the drain.

Inorganic halides	Pour onto a DRY 50-50 mixture of kaolin and sodium carbonate. While stirring, slowly add 6 M ammonium hydroxide. Add ice or cold water and stir until ammonium chloride smoke is no longer produced. Neutralize with 6 M HCl and flush down drain.
Inorganic peroxides	Cover with 90:10% sand/sodium carbonate mixture. Slowly add this mixture to a large volume of sodium sulfite solution. Neutralize with 3 M sulfuric acid. Flush the liquid down the drain; send solid (sand) to a landfill.
Inorganic sulfides	Add to a ferric chloride solution. Add sodium carbonate until neutral. Flush down drain.
Inorganic salts (other than previously listed)	Dissolve in a large volume of water. Add sodium carbonate and let the solution stand overnight. Neutralize with 6 M HCl and flush down drain.
Nonmetal compounds	Sift into 50-50 mixture of sodium carbonate and calcium hydroxide. Spray with water, then add to large volume of water. Neutralize with either 6 M HCl or 6 M ammonium hydroxide. Flush down drain.
Oxidizing agents	Add to a large volume of water containing a bisulfite or hypo as a reducing agent. Neutralize, and flush down drain.
Phosphorus	
Red	Burn in a hood or open area.
Yellow	Cover with water and place in a hood or open area. When water evaporates, dry phosphorus will ignite and burn.
Reducing substances	Cover with sodium carbonate and spray with water. Slowly add equal volume of calcium hypochlorite, stir, and add more water. Let stand overnight, neutralize, and flush down drain.

Organic Chemicals

Chemical	**Disposal Procedure**
Alcohols	Absorb on paper. Evaporate in a hood or open area. Burn the paper.
Aldehydes	Absorb on paper. Burn in a hood or open area.
Aliphatic amines	Add to solid sodium bisulfate in a large dish. Add water and neutralize. Flush down drain.
Aromatic amines	Pour onto 90:10% sand/sodium carbonate mixture. Place in cardboard box, fill with paper, and burn in incinerator.
Aromatic halogenated amines	Same as the Aromatic amines
Aromatic nitro compounds	Same as for Aromatic amines
Esters	Same as for Alcohols
Ethers	Pour on ground in open area. Ignite from a safe distance.
Hydrocarbons	Same as for Alcohols
Ketones	Same as for Alcohols

Organic acids	Nonvolatile acids can be mixed with sodium carbonate, neutralized, and flushed down drain. Volatile acids can be mixed with flammable solvent and burned.
Organic acid halides	Add to 50-50 mixture of sodium carbonate-calcium hydroxide. Slowly add 6 M ammonium hydroxide. Dilute to large volume. Neutralize and flush down drain.
Organic halogens	Same as for Aromatic amines
Organic peroxides	Place in plastic container and burn in an open area or incinerator. Or, add to 20% NaOH solution. Let the solution sit overnight, neutralize, and flush down drain.
Organic phosphates	Mix with equal parts pulverized limestone and sand. Wet with alcohol or benzene, and burn in open incinerator.
Substituted organic acids	Same as for Organic acid halides

Index

A

Acetylene production, 16
Acid-base indicators
  effect of carbon dioxide, 40
  general, 37
  pH, 39
  universal indicator, 38
Activity series for some metals, 103
Ammonia, catalytic oxidation, 87
Ammonia fountain, solubility of a gas, 10–11
Ammonium hydroxide-ammonia system, effect of temperature on equilibrium, 62
Aniline, oxidation products, 169
Atomic diameter, properties of atoms, 3
Atomic properties, electronegativity, atomic diameter, and ionization energy, 3
Autocatalysis, kinetics, 73–74

B

Battery, electrochemistry, 117–118
Bleach, laundry
  preparation of chlorine gas, 13
  preparation of oxygen gas, 12
Boiling point, effect of pressure on gases, 19

C

Carbon as a catalyst, kinetics, 69
Carbon dioxide, effect on acid-base indicators, 40
Catalysis, kinetics, 72
Catalyst, carbon, kinetics, 69
Catalytic decomposition, kinetics, 71
Chemiluminescence, luminol and various oxidizers, 138, 140–141
Chlorine gas, preparation from laundry bleach, 13
Chromate-dichromate system, equilibrium, 58–59
Cobalt complex, effect of concentration on equilibrium, 56–57
Colloidal system
  production of a foam, 135, 136
  production of a gel, 134
  sulfur, a chemical sunset, 132–133
Common ion effect
  equilibrium, 159
  reaction rate, 154, 155–156
  solubility, 157–158
Complex ion, red, white, and blue solutions, 148
Concentration
  effect on equilibrium
    chromate-dichromate system, 58–59
    cobalt complex, 56–57
    copper complex, 54–55
  effect on solubility products, 66
Coordination numbers, effect on solubility and solutions, 32
Copper complex, effects of concentration and temperature on equilibrium, 54–55
Crystal formation
  metal trees, 153
  solubility and solutions, 28–29

D

Decomposition, catalytic, 71
Dehydration
  p-nitroaniline, 125
  sucrose, 124

Densities, gases, 7
Diffusion, gases, 14-15
Diphosphorus tetrahydride, production and ignition, 171-172
Displacement reaction
  formation of a metal, 153
  tin by zinc, electrochemistry, 119
Double displacement reaction
  between two solids, 123
  red, white, and blue solutions, 148

E

Electrochemistry
  displacement of tin by zinc, 119
  simple battery, 117-118
Electronegativity, properties of atoms, 3
Endothermic reaction, energy changes, 43, 44, 45
Energy changes
  endothermic reactions, 43, 44, 45
  exothermic reactions, 46, 47
Equilibrium
  common ion effect, 159
  effect of concentration
    chromate-dichromate system, 58-59
    cobalt complex, 56-57
    copper complex, 54-55
    solubility products, 66
  effect of hydrolysis, 65
  effect of pressure, 64
  effect of temperature, 53, 62, 63
  LeChatelier's Principle, 51-52
Exothermic reaction
  energy changes, 46, 47
  production and spontaneous combustion of acetylene, 164-165
Explosion, solid-solid reaction, 170
Explosive
  combination of hydrogen and oxygen, 175-176
  simple, nitrogen triiodide, 173-174

F

Foam, production, 135

G

Gases
  chlorine, preparation from laundry bleach, 13
  densities, 7
  determining molecular weight, 17-18
  diffusion, 14-15
  effect of pressure on boiling point, 19
  effect of temperature on equilibrium, 60
  oxygen, preparation from laundry bleach, 12
  production, acetylene, 16
  properties, pressure and suction, 8
  solubility, ammonia fountain, 10
  temperature and pressure relationships, 9
Gel formation, production of sterno, 134
Glycerin, oxidation by permanganate, 88

H

Hydrate, effect of temperature on solubility and solutions, 31
Hydrogen and oxygen, explosive combination, 175-176
Hydrolysis, effect on equilibrium, 65

I

Ignition of diphosphorus tetrahydride, 171-172
Indicators, acid-base
  effect of carbon dioxide, 40
  and pH, 39
  various, 37
Ionization energy, properties of atoms, 3

K

Kinetics
  autocatalysis, 73-74
  carbon as a catalyst, 69
  catalysis, 72
  catalytic decomposition of hydrogen peroxide, 71

Kinetics—*Continued*
Old Nassau reaction, 77
oscillating reactions, 81–83
quick gold reaction, 80
solubility and complex formation, 78
starch-iodine reaction, 75–76
temperature effect, 70
traffic light reaction, 79

L

LeChatelier's Principle, equilibrium, 51–52

M

Manganese, oxidation states, 98
Metallurgy, 104
Molecular weight determination, gases, 17–18

N

Negative coefficient of solubility, 27
p-Nitroaniline, dehydration, 125
Nitrogen triiodide, a simple explosive, 173–174
Nylon, synthesis, 126–127

O

Oscillating reaction, kinetics, 81, 82, 83
Oxidation products, aniline, 169
Oxidation reactions
alcohol by potassium permanganate, 106–107
catalytic, of ammonia, 87
color changes in Fe(II) and Fe(III) solutions, 143–144
decomposition of ammonium dichromate, 142
glycerin by permanganate, 88
phosphorus, 166–168
sodium, 105
sugar, decomposition with sulfuric acid, 163
zinc, 95
Oxidation-reduction reactions
activity series for some metals, 103
blue bottle reaction, 93–94
blueprint reaction, 113–114
copper into gold, 104
formation of a metal, 152
iron, 89–90
iron oxalate, 145–146
mercury beating heart, 96–97
Prussian blue reaction, 101–102
silver mirror reaction, 91–92
Oxidation states
chromium, 112
manganese, 98, 99–100
vanadium, 108–109, 110–111
Oxidizers, various, and luminol, chemiluminescence, 139–140
Oxidizing agent
chemiluminescence with luminol, 138
strong, benzoyl peroxide, 169
Oxygen gas
explosive combination with hydrogen, 175–176
preparation from laundry bleach, 12

P

Permanganate, oxidation of glycerin, 88
pH and acid-base indicators, 39
Phosphine, production and ignition, 171–172
Phosphorus, oxidation, 166–167
Photoreduction, 113–114
Polar properties and solubility, 33–34
Pressure
effect on boiling point of gases, 19
effect on equilibrium, 64
effect on properties of gases, 8
effect on solid-solid reaction, 170
Pressure relationships, gases, 9

R

Rayon, synthesis, 128–129
Reaction rate, common ion effect, 154, 155–157
Redox reactions, writing reactions, 147

Reoxidation, vanadium, 110-111
Rubber, synthesis, 130-131

S

Solid-solid reaction
double displacement, 123
initiated by pressure, 170
Solubility
common ion effect, 157-158
crystallization, 30
effect of changing coordination numbers, 32
effect of temperature, 26
effect of temperature on a hydrate, 31
Fe(II) and Fe(III) solutions, 143-144
gas, the ammonia fountain, 10
negative coefficient, 27
polar properties, 33-34
precipitate formation, 23-25
product, effect of concentration, 66
supersaturation and crystallization, 28-29
Solutions
crystallization, 30
effect of changing coordination numbers, 32
effect of temperature on a hydrate, 31
precipitate formation, 23-25
supersaturation and crystallization, 28-29
Structural stability, explosives, 173-174
Sucrose, dehydration, 124
Suction, gases, properties, 8
Supersaturation, solubility and solutions, 28-29
Surface area, effect in chemical reactions, 123
Surface tension of water, 137
Synthesis
nylon, 126-127
rayon, 128-129
rubber, 130-131

T

Temperature
effect on a hydrate, solubility and solutions, 31
effect on equilibrium, 53, 62, 63
copper complex, 54-55
gases, 60
effect on kinetics, 70
effect on solubility, 26
Temperature relationships, gases, 9

U

Universal indicator, acid-base indicators, 38

W

Water, surface tension, 137
Wetting agent, surface tension of water, 137

Z

Zinc, oxidation, 95